Invariant Differential Operators for Quantum Symmetric Spaces

MEMOIRS
of the
American Mathematical Society

Number 903

Invariant Differential Operators for Quantum Symmetric Spaces

Gail Letzter

May 2008 • Volume 193 • Number 903 (third of 5 numbers) • ISSN 0065-9266

American Mathematical Society
Providence, Rhode Island

2000 *Mathematics Subject Classification.*
Primary 17B37.

Library of Congress Cataloging-in-Publication Data

Letzter, Gail, 1960–
Invariant differential operators for quantum symmetric spaces / Gail Letzter.
p. cm. — (Memoirs of the American Mathematical Society, ISSN 0065-9266 ; no. 903)
"May 2008, volume 193, number 903 (third of 5 numbers)."
Includes bibliographical references.
ISBN 978-0-8218-4131-0 (alk. paper)
1. Quantum groups. I. Title.

QC20.7.G76L48 2008
512′.482—dc22 2008060004

Memoirs of the American Mathematical Society

This journal is devoted entirely to research in pure and applied mathematics.

Subscription information. The 2008 subscription begins with volume 191 and consists of six mailings, each containing one or more numbers. Subscription prices for 2008 are US$675 list, US$540 institutional member. A late charge of 10% of the subscription price will be imposed on orders received from nonmembers after January 1 of the subscription year. Subscribers outside the United States and India must pay a postage surcharge of US$38; subscribers in India must pay a postage surcharge of US$43. Expedited delivery to destinations in North America US$53; elsewhere US$130. Each number may be ordered separately; *please specify number* when ordering an individual number. For prices and titles of recently released numbers, see the New Publications sections of the *Notices of the American Mathematical Society.*

Back number information. For back issues see the *AMS Catalog of Publications.*

Subscriptions and orders should be addressed to the American Mathematical Society, P. O. Box 845904, Boston, MA 02284-5904, USA. *All orders must be accompanied by payment.* Other correspondence should be addressed to 201 Charles Street, Providence, RI 02904-2294, USA.

Memoirs of the American Mathematical Society (ISSN 0065-9266) is published bimonthly (each volume consisting usually of more than one number) by the American Mathematical Society at 201 Charles Street, Providence, RI 02904-2294, USA. Periodicals postage paid at Providence, RI. Postmaster: Send address changes to Memoirs, American Mathematical Society, 201 Charles Street, Providence, RI 02904-2294, USA.

This publication is indexed in *Science Citation Index*®, *SciSearch*®, *Research Alert*®, *CompuMath Citation Index*®, *Current Contents*®*/Physical, Chemical & Earth Sciences.*
Printed in the United States of America.

∞ The paper used in this book is acid-free and falls within the guidelines established to ensure permanence and durability.
Visit the AMS home page at `http://www.ams.org/`

10 9 8 7 6 5 4 3 2 1 13 12 11 10 09 08

Contents

Abstract

This paper studies quantum invariant differential operators for quantum symmetric spaces in the maximally split case. The main results are quantum versions of theorems of Harish-Chandra and Helgason: There is a Harish-Chandra map which induces an isomorphism between the ring of quantum invariant differential operators and the ring of invariants of a certain Laurent polynomial ring under an action of the restricted Weyl group. Moreover, the image of the center under this map is the entire invariant ring if and only if the underlying irreducible symmetric pair is not of four exceptional types. In the process, we find a particularly nice basis for the quantum invariant differential operators that provides a new interpretation of difference operators associated to Macdonald polynomials.

Received by the editor January 12, 2005.

2000 *Mathematics Subject Classification.* Primary 17B37 .

Key words and phrases. quantized enveloping algebras, quantum symmetric spaces, Macdonald polynomials.

Supported by grants from NSA.

Introduction

Harmonic analysis on symmetric spaces studies invariant differential operators and their joint eigenspaces in connection with Lie groups. The discovery of quantum groups in the 1980's inspired the growing subject of harmonic analysis on quantum symmetric spaces. In particular, zonal spherical functions on most compact quantum symmetric spaces have been identified with Macdonald or Macdonald-Koornwinder polynomials ([N], [NS], [S], [L5], [DN], [NDS], and [DS]). In this paper, the focus is on quantum invariant differential operators. We prove a quantum analog of Harish-Chandra's fundamental result: the Harish-Chandra map induces an isomorphism between the ring of invariant differential operators on a symmetric space and invariants of an appropriate polynomial ring under the restricted Weyl group. We further establish a quantum version of a related theorem due to Helgason: the image of the center under this Harish-Chandra map is the entire invariant ring if and only if the underlying irreducible symmetric pair is not one of four exceptional types, EIII, EIV, EVII, or EIX. Moreover, we exhibit a nice basis for the quantum invariant differential operators which corresponds to a special commuting family of difference operators associated to Macdonald polynomials.

Let $\mathfrak{g}$ be a semisimple complex Lie algebra. A classical (infinitesimal) symmetric pair is a pair of Lie algebras $\mathfrak{g}, \mathfrak{g}^\theta$ where $\mathfrak{g}^\theta$ is the Lie subalgebra fixed by an involution θ. One can associate an infinitesimal symmetric pair to each symmetric space. In this algebraic framework, invariant differential operators on the symmetric space correspond to ad $\mathfrak{g}^\theta$ invariant elements of $U(\mathfrak{g})$. Thus a quantum invariant differential operator should be an element of the quantum analog of the fixed ring of $U(\mathfrak{g})$ under the action of $\mathfrak{g}^\theta$. More precisely, let $\check{U}$ denote the simply connected quantized enveloping algebra of $\mathfrak{g}$ over the algebraic closure $\mathcal{C}$ of $\mathbf{C}(q)$. Quantum symmetric pairs are defined using left coideal subalgebras B of $\check{U}$ which can be viewed as quantum analogs of the enveloping algebra of $\mathfrak{g}^\theta$ ([L3, Section 7]). Our investigation of quantum invariant differential operators is an analysis of the ring $\check{U}^B$ of invariants in $\check{U}$ with respect to the right adjoint action of B. In analogy to the classical case, quantum zonal spherical functions are eigenvectors with respect to the action of $\check{U}^B$ on the quantized function algebra associated to $\mathfrak{g}$ (see Theorem 1.4).

Let $\mathfrak{a}$ denote the eigenspace for θ with eigenvalue -1 inside a maximally split Cartan subalgebra $\mathfrak{h}$ of $\mathfrak{g}$. In the classical case, the Harish-Chandra map associated to $\mathfrak{g}, \mathfrak{g}^\theta$ is the projection from $U(\mathfrak{g})$ onto the enveloping algebra of $\mathfrak{a}$, defined using the Iwasawa decomposition of $\mathfrak{g}$. This picture can be lifted to the quantum setting as follows. The Cartan subalgebra $\check{U}^0$ of $\check{U}$ is the group algebra of a multiplicative group $\check{T}$ isomorphic to the weight lattice of $\mathfrak{g}$. The restricted root system Σ associated to the pair $\mathfrak{g}, \mathfrak{g}^\theta$ spans the vector space $\mathfrak{a}^*$. A quantum analog for $\mathfrak{a}$ is a multiplicative subgroup $\mathcal{A}$ of $\check{T}$ isomorphic to the weight lattice of 2Σ. Using a quantum version of the Iwasawa decomposition (Theorem 4.2 below), the quantum Harish-Chandra map $\mathcal{P}_B$ is a projection of $\check{U}$ onto a slight extension of $\mathcal{C}[\mathcal{A}]$. The map $\mathcal{P}_B$ can be used to determine the eigenvalues of these zonal spherical functions with respect to the action of $\check{U}^B$ (Lemma 5.3).

The action of the restricted Weyl group W_Θ on Σ induces an action of W_Θ on $\mathcal{C}[\mathcal{A}]$. This action can be further twisted to a dotted (or translated) action of W_Θ on $\mathcal{C}[\mathcal{A}]$. The first main result of this paper is:

Theorem A: *The Harish-Chandra map $\mathcal{P}_B$ induces a surjection from $\check{U}^B$ onto the subring of invariants of $\mathcal{C}[\mathcal{A}]$ with respect to the dotted action of W_Θ.*

Perhaps the most intricate part in the proof of the above theorem is showing that $\mathcal{P}_B(\check{U}^B)$ is invariant under the dotted action of W_Θ. Recall that there is no direct quantum version of the Lie group G associated to $\mathfrak{g}$; there is only a quantum analog of the ring of regular functions on G. Thus the analytic techniques used in Harish-Chandra's original work [HC] are unavailable here. Lepowsky's algebraic proof [Le] using the Cartan decomposition of $\mathfrak{g}$ with respect to $\mathfrak{g}^\theta$ cannot be adapted to the quantum setting because there is no obvious quantum Cartan decomposition. Let $Z(\check{U})$ denote the center of $\check{U}$ and note that $Z(\check{U})$ is a subalgebra of $\check{U}^B$. Ultimately, the dotted W_Θ invariance of $\mathcal{P}_B(\check{U}^B)$ follows from the dotted W_Θ invariance of $\mathcal{P}_B(Z(\check{U}))$. Indeed, for all but four types of symmetric pairs, Theorem A is a consequence of the following quantum version of Helgason's theorem.

Theorem B: *The image of $Z(\check{U})$ under $\mathcal{P}_B$ is equal to the subring of invariants of $\mathcal{C}[\mathcal{A}]$ with respect to the dotted action of W_Θ if and only if $\mathfrak{g}, \mathfrak{g}^\theta$ is not of type EIII, EIV, EVII, or EIX.*

It turns out that the restriction of $\mathcal{P}_B$ to $Z(\check{U})$ is just the composition of the ordinary quantum Harish-Chandra map $\mathcal{P}$ followed by restriction of $\mathcal{P}_B$ to $\check{U}^0$. It is well known that $\mathcal{P}(Z(\check{U}))$ consists of the invariants of a particular Laurent polynomial subring of $\check{U}^0$ with respect to the dotted action of the large Weyl group. Thus determining $\mathcal{P}_B(Z(\check{U}))$ reduces to computing the image of the invariants of this Laurent polynomial ring inside $\mathcal{C}[\mathcal{A}]$. When $\mathfrak{g}, \mathfrak{g}^\theta$ is one of the four types EIII, EIV, EVII, EIX, we show that $\mathcal{P}_B(Z(\check{U}))$ is not equal to the dotted W_Θ invariants of $\mathcal{C}[\mathcal{A}]$ using specialization at $q = 1$ and the corresponding classical results. Unfortunately, specialization fails to show the other direction of the primary assertion of the theorem. Instead, we use information relating the weight lattice of the root system of $\mathfrak{g}$ to that of Σ combined with properties of a particular generating set of $Z(\check{U})$ to determine $\mathcal{P}_B(Z(\check{U}))$. The proof turns out to be more delicate for those symmetric pairs related to the irreducible symmetric pair of type AII. In Section 8, we use combinatorial and character formula arguments to handle this family.

We take a different approach in order to establish Theorem A for the remaining four types of irreducible symmetric pairs. In particular, in Sections 7,8, and 9, we find a nice basis for the vector space $\mathcal{P}_B(\check{U}^B)$. Let $F_r(\check{U})$ denote the locally finite part of $\check{U}$ with respect to the right adjoint action. We construct a projection map $\mathcal{L}$ from $F_r(\check{U})$ onto $\check{U}^B$ that can be viewed as a generalization of the Reynolds operator. Write τ for the isomorphism between $\mathcal{A}$ and the set $\{2\lambda|\lambda$ is in the weight lattice of $\Sigma\}$ and let $P^+(\Sigma)$ denote the semigroup of dominant integral restricted weights associated to Σ. We show that Theorem A is a consequence of fine information concerning the special basis of $\mathcal{P}_B(\check{U}^B)$ obtained in the following proposition.

Theorem C: *For each $\lambda \in P^+(\Sigma)$, there exists an element $t_{2\lambda}$ in the intersection of $\tau(2\lambda) + \text{Ker } \mathcal{P}_B$ and $F_r(\check{U})$ such that $\{\mathcal{P}_B(\mathcal{L}(t_{2\lambda}))|\lambda \in P^+(\Sigma)\}$ forms a basis for $\mathcal{P}_B(\check{U}^B)$. Moreover there is a nice formula (see Lemma 7.5) relating the evaluation of a quantum zonal spherical function at $\mathcal{L}(t_{2\lambda})$ and the image of $\mathcal{L}(t_{2\lambda})$ under $\mathcal{P}_B$.*

The original motivation for Theorem C was to show that Theorem A holds for the remaining four exceptional types. However, Theorem C, which appears quite technical, has significance which goes beyond the particulars of this paper. We will ultimately use Theorem C to develop the theory of noncompact zonal spherical functions. Recall that compact quantum zonal spherical functions are elements of the quantized function algebra of the compact Lie group G associated to $\mathfrak{g}$. Unfortunately, there is not yet a good definition of the quantized function algebra associated to a noncompact semisimple Lie group. As a result, the only noncompact zonal spherical functions that have been analyzed so far are those on the simplest noncompact symmetric space associated to $\mathfrak{sl}\ 2$ ([KS]). In a future paper, Theorem C will be the foundation of a general algebraic definition of zonal spherical functions. The general idea is as follows. Set $\mathcal{K} = (\text{ad}_r\ B_+)\check{U} + \text{Ker } \mathcal{P}_B$. Theorem C and the definition of $\mathcal{L}$ guarantee that

$$\check{U}^B \oplus \mathcal{K} = \mathcal{C}[\tau(2\lambda)|\ \lambda \in P^+(\Sigma)] \oplus \mathcal{K}$$

where both sides of this equality are subspaces of $\check{U}$. Using the definition of the projection map $\mathcal{L}$, one further shows that $F_r(\check{U})$ is a subset of this direct sum of vector spaces. Given an algebra homomorphism Λ from $\mathcal{C}[\mathcal{A}]$ to $\mathcal{C}$, define a linear function g_Λ on $\check{U}^B$ by $g_\Lambda(a) = \Lambda(\mathcal{P}_B(a))$ for $a \in \check{U}^B$. Then g_Λ extends to a function on the direct sum (and hence on $F_r(\check{U})$) where the restriction of g_Λ to $\mathcal{K}$ is zero. Using the basic local finiteness theorem ([JL1, Theorem 6.4]), g_Λ extends to a function on $\check{U}$ for good choices of Λ. For each restricted integral weight γ, we write z^γ for the algebra homomorphism from $\mathcal{C}[\mathcal{A}]$ to $\mathcal{C}$ that sends $\tau(2\lambda)$ to $q^{(2\lambda,\gamma)}$ for each λ. It turns out that if Λ is of the form $z^{2\mu}$ for $\mu \in P^+(\Sigma)$ then g_Λ is a compact quantum zonal spherical function at 2μ associated to B. Noncompact quantum zonal spherical functions correspond to choices of Λ that are not of the form $z^{2\mu}$ for $\mu \in P^+(\Sigma)$.

The reader may wish to focus instead on completing the identification of compact quantum zonal spherical functions with orthogonal polynomials. Indeed, compact quantum zonal spherical functions associated to standard quantum symmetric pairs with reduced restricted root systems are identified with Macdonald polynomials in [L5] by computing radial components of "small" elements in $\check{U}^B$. For most irreducible symmetric pairs, these small elements are actually contained in the center $Z(\check{U})$. However, this is not true for the four problematic exceptional types EIII, EIV, EVII, and EIX. In [L5], an elementwise computation was used to find the necessary elements of $\check{U}^B$ for the last three of these types. The arguments of this paper provide a simpler proof of their existence. Moreover, Theorem A guarantees the existence of B invariant elements necessary to make the methods of [L5] extend to the remaining problematic type, EIII.

There is another function on $\check{U}$, the radial component map $\mathcal{X}$, which is closely related to $\mathcal{P}_B$ and is useful in analyzing $\mathcal{P}_B(\check{U}^B)$. We show that the map which sends $\mathcal{P}_B(u)$ to $\mathcal{X}(u)$ defines an algebra isomorphism of $\mathcal{P}_B(\check{U}^B)$ onto $\mathcal{X}(\check{U}^B)$. This enables us to determine the kernel of $\mathcal{P}_B(\check{U}^B)$. It should be noted that the quotient of $\check{U}^B$ modulo this kernel is the exact quantum analog of the ring of invariant differential operators on the corresponding symmetric space. Furthermore, Theorem A and Theorem 5.6 ensure that $\mathcal{X}(\check{U}^B)$ is a polynomial ring in rank Σ variables. With the help of $\mathcal{X}$, we interpret elements of $\check{U}^B$ as difference operators acting on the character group ring of Σ. It is precisely this interpretation that leads to the identification of quantum zonal spherical functions as Macdonald (or Macdonald-Koornwinder) polynomials (see for example [N], [NS], [L5], [DN], [NDS], and [DS]). Theorem A ensures that $\mathcal{X}(\check{U}^B)$ forms a completely integrable system of difference operators whose eigenfunctions are the orthogonal polynomials associated to quantum zonal spherical functions. Moreover, the basis for $\check{U}^B$ described in Theorem C translates via $\mathcal{X}$ to a special basis for the commuting family of difference operators associated to Macdonald polynomials and related orthogonal polynomials (see the discussion following Theorem 7.9 and [K, Theorem 6.6]).

We describe the organization of this paper. Section 1 sets notation and reviews basic background concerning classical symmetric pairs and related root systems, quantized enveloping algebras, quantum symmetric pairs, and quantum compact zonal spherical functions.

Section 2 and Section 3 focus on properties of the root systems and related Weyl groups associated to the symmetric pair $\mathfrak{g}, \mathfrak{g}^\theta$. Section 2 is an analysis of the image of the weight lattice of the root system associated to the underlying semisimple Lie algebra $\mathfrak{g}$ with respect to the map $\tilde{}$ defined by restriction to $\mathfrak{a}$. In Section 3, we study connections between the dotted actions of both the large Weyl group and restricted Weyl group on subrings of the Cartan subalgebra of the quantized enveloping algebra. This information is later applied to show that the image of the center of the quantized enveloping algebra under the Harish-Chandra map $\mathcal{P}_B$ is invariant under the dotted action of the restricted Weyl group.

In Sections 4 and 5, we turn to the setting of quantum symmetric pairs. The two maps used to study the quantum invariant differential operators are defined based on similar work in [L5]. In Section 4, we extend the quantum Iwasawa decomposition of [L5] to all quantum symmetric pairs and use it to construct the Harish-Chandra map $\mathcal{P}_B$. The definition of the quantum radial component map $\mathcal{X}$ as described in [L5] is extended to the general case in Section 5. Connections between $\mathcal{P}_B(\check{U}^B)$ and $\mathcal{X}(\check{U}^B)$ are established, leading to a nice description of the possible highest degree terms of elements in $\mathcal{P}_B(\check{U}^B)$.

Sections 6 lays the ground work for the proof of Theorem B while Section 7 provides the theory necessary to prove Theorem C. Section 6 analyzes the image of the center under $\mathcal{P}_B$ using information about the image of $Z(\check{U})$ under the ordinary Harish-Chandra map. In Section 7, we introduce the map $\mathcal{L}$ and prove special properties of elements of the form $\mathcal{P}_B(\mathcal{L}(a))$ for $a \in F_r(\check{U})$. A simple formula relating the value of a zonal spherical function at an element a in $F_r(\check{U})$ and $\mathcal{P}_B(\mathcal{L}(a))$ is presented in Lemma 7.5. The linear independence of $\{\mathcal{P}_B(\mathcal{L}(t_{2\lambda}))|\lambda \in P^+(\Sigma)\}$ is a consequence of this formula (Theorem 7.7). A study of the possible highest degree term of $\mathcal{L}(a)$ for any a in the set $F_r(\check{U}) \cap (\tau(2\lambda) + \text{Ker } \mathcal{P}_B)$ (see Lemma 7.6 and Theorem 7.7) is ultimately used to establish the spanning part of Theorem C.

Throughout much of the paper, we have kept the case-by-case analysis to a mininum. Indeed, by the end of Section 7, Theorems A, B, and C have been established for all irreducible symmetric pairs except for two different families. However, case work is necessary to complete the proofs of Theorems A,B, and C. In particular, the last two sections of the paper are each devoted to one of these two families. Section 8 completes the proofs of Theorems A, B, and C for those irreducible symmetric pairs that contain an irreducible symmetric pair of type AII. Section 9 handles the four exceptional irreducible symmetric pairs, EIII, EIV, EVII, and EIX. In should be noted that in both sections, the proofs do not rely on element manipulation. Instead, we reduce the proofs to computations in the commutative polynomial ring of dotted W_Θ invariants inside $\mathcal{C}[\mathcal{A}]$ which resemble character formula arguments.

A list of symbols defined in Section 1 followed by the definition of symbols introduced after Section 1 can be found in an appendix at the end of the paper.

Acknowledgements. The author would like to thank Nolan Wallach for his wonderful insight and suggestions which led to this project and Dan Farkas for his helpful comments on writing. The author would also like to thank the referee for conscientiously reading an earlier version of this paper and providing many helpful suggestions.

CHAPTER 1

Background and Notation

Let $\mathbf{C}$ denote the complex numbers, $\mathbf{Q}$ denote the rational numbers, $\mathbf{Z}$ denote the integers, $\mathbf{R}$ denote the real numbers, $\mathbf{N}$ denote the nonnegative integers, and q denote an indeterminate. Let $\{q^r \mid r \in \mathbf{Q}\}$ denote the multiplicative group isomorphic to $\mathbf{Q}$ under addition. Write $\mathbf{C}(\mathbf{Q})$ for the quotient field of the group algebra of $\{q^r \mid r \in \mathbf{Q}\}$ over $\mathbf{C}$. Similarly, let $\mathbf{R}(\mathbf{Q})$ denote the quotient field of the group algebra of $\{q^r \mid r \in \mathbf{Q}\}$ over $\mathbf{R}$. Note that $\mathbf{R}(\mathbf{Q})$ can be made into an ordered field in a manner similar to the rational function field $\mathbf{R}(q)$ (see [Ja, Section 5.1]) Write $\mathcal{C}$ for the algebraic closure of $\mathbf{C}(\mathbf{Q})$ and let $\mathcal{R}$ denote the real algebraic closure of $\mathbf{R}(\mathbf{Q})$.

Suppose that Φ is a root system. Write $Q(\Phi)$ for the root lattice of Φ and let $Q^+(\Phi)$ be the subset of $Q(\Phi)$ equal to the $\mathbf{N}$ span of the positive roots in Φ. Let $P(\Phi)$ denote the weight lattice of Φ and let $P^+(\Phi)$ be the subset of $P(\Phi)$ consisting of dominant integral weights. (Sometimes we replace Φ with the symbol used to represent the positive simple roots in the notation for the weight and root lattices and their subsets.)

Let $\mathfrak{g}$ be a complex semisimple Lie algebra with triangular decomposition $\mathfrak{g} = \mathfrak{n}^- \oplus \mathfrak{h} \oplus \mathfrak{n}^+$. Let Δ denote the root system of $\mathfrak{g}$ and let $(\ ,\ )$ denote the corresponding Cartan inner product on Δ. Here we assume that the positive roots of Δ correspond to the root vectors in $\mathfrak{n}^+$. Write $\pi = \{\alpha_1, \ldots, \alpha_n\}$ for the positive simple roots of Δ. We use "$\leq$" to denote the usual partial order on $\mathfrak{h}^*$. In particular, given α and β in $\mathfrak{h}^*$, we say that $\alpha \leq \beta$ if and only if $\beta - \alpha \in Q^+(\pi)$. Let $\{e_i, h_i, f_i | 1 \leq i \leq n\}$ be a standard set of generators for $\mathfrak{g}$. Here, e_i is a root vector in $\mathfrak{n}^+$ corresponding to the simple root α_i, f_i is a root vector in $\mathfrak{n}^-$ corresponding to the root $-\alpha_i$, and $h_1, \ldots, h_n$ is a basis of coroots for $\mathfrak{h}$.

Classical (infinitesimal) symmetric pairs: Let θ be a Lie algebra involution on $\mathfrak{g}$ and write $\mathfrak{g}^\theta$ for the fixed Lie subalgebra of $\mathfrak{g}$ with respect to θ. The pair $\mathfrak{g}, \mathfrak{g}^\theta$ is a classical (infinitesimal) symmetric pair. We assume throughout the paper that $\mathfrak{g}, \mathfrak{g}^\theta$ is an irreducible symmetric pair. The results of this paper extend in a straightforward manner to the general case. More precisely, a symmetric pair $\mathfrak{g}, \mathfrak{g}^\theta$ is called irreducible provided that $\mathfrak{g}$ cannot be written as a direct sum of semisimple Lie algebras which both admit θ as an involution. The classification of involutions and classical irreducible symmetric pairs up to isomorphism can be found in [A] (see also [L4, Section 7].)

Recall that we have already specified a particular Cartan subalgebra $\mathfrak{h}$ of $\mathfrak{g}$. We assume that θ is a maximally split involution with respect to $\mathfrak{h}$ in the sense of [L3, Section 7]. In particular, θ is *maximally split* with respect to $\mathfrak{h}$ provided the following three conditions hold.

(1.1) $\theta(\mathfrak{h}) = \mathfrak{h}$

(1.2) If $\theta(h_i) = h_i$ then $\theta(e_i) = e_i$ and $\theta(f_i) = f_i$

(1.3) If $\theta(h_i) \neq h_i$, then $\theta(e_i)$ is a nonzero root vector in $\mathfrak{n}^-$ and $\theta(f_i)$ is a nonzero root vector in $\mathfrak{n}^+$.

It should be noted that in the classical case, one can always choose a Cartan subalgebra so that it is maximally split with respect to a particular involution of $\mathfrak{g}$. Alternatively, given an involution θ' of $\mathfrak{g}$, one can find a Lie algebra automorphism ψ of $\mathfrak{g}$ so that $\psi\theta'\psi^{-1}$ satisfies the above three conditions with respect to a prechosen Cartan subalgebra $\mathfrak{h}$ of $\mathfrak{g}$ ([D, Section 1.13]). Unfortunately, such flexibility does not exist in the quantum case. We only consider here lifts of symmetric pairs to the quantum case using maximally split involutions with respect to $\mathfrak{h}$.

There is a second root system associated to the pair $\mathfrak{g}, \mathfrak{g}^\theta$, referred to as the set of restricted roots Σ, defined using the involution θ. More precisely, (1.1), (1.2), and (1.3) ensure that θ induces an involution Θ on $\mathfrak{h}^*$ which restricts to an involution on Δ. Given $\alpha \in \mathfrak{h}^*$, set

$$\tilde{\alpha} = (\alpha - \Theta(\alpha))/2. \tag{1.4}$$

The restricted root system Σ is the set

$$\Sigma = \{\tilde{\alpha} | \alpha \in \Delta \text{ and } \Theta(\alpha) \neq \alpha\}.$$

Moreover, Σ inherits the structure of a root system using the inner product of Δ ([Kn, Chapter VI, Section 4]). Set $\pi_\Theta = \{\alpha_i \in \pi | \Theta(\alpha_i) = \alpha_i\}$. Recall [L3, Section 7, (7.5)] that there exists a permutation p of the set $\{1, 2, \ldots, n\}$ such that p induces a diagram automorphism on π and

$$\Theta(-\alpha_i) - \alpha_{\mathrm{p}(i)} \in Q^+(\pi_\Theta) \tag{1.5}$$

for $\alpha_i \in \pi \setminus \pi_\Theta$. Set $\pi^* = \{\alpha_i \in \pi \setminus \pi_\Theta | \, i \leq \mathrm{p}(i)\}$. The set $\{\tilde{\alpha}_i | \, \alpha_i \in \pi^*\}$ is the set of positive simple roots for the root system Σ.

Quantized enveloping algebras: Let $q_i = q^{(\alpha_i,\alpha_i)/2}$. Set

$$[m]_q = (q^m - q^{-m})/(q - q^{-1}) \quad \text{and} \quad [m]_q! = [m]_q[m-1]_q \cdots [1]_q.$$

Define the q binomial coefficients by

$$\begin{bmatrix} m \\ j \end{bmatrix}_q = \frac{[m]!_q}{[j]!_q[m-j]!_q}.$$

Set $a_{ij} = 2(\alpha_i, \alpha_j)/(\alpha_i, \alpha_i)$ for $1 \leq i \leq n$ and $1 \leq j \leq n$. Let $U = U_q(\mathfrak{g})$ denote the quantized enveloping algebra of $\mathfrak{g}$. The algebra U is generated over $\mathcal{C}$ by $x_i, y_i, t_i^{\pm 1}$, $1 \leq i \leq n$, subject to the following relations (see for example [L3, Section 1, (1.4)-(1.10)], [Jo, 3.2.9], or [DK, Section 1]).

(1.6) $x_iy_j - y_jx_i = \delta_{ij}\frac{(t_i - t_i^{-1})}{(q_i - q_i^{-1})}$ for each $1 \leq i \leq n$.

(1.7) The $t_1^{\pm 1}, \ldots, t_n^{\pm 1}$ generate a free abelian group T of rank n.

(1.8) $t_ix_j = q^{(\alpha_i,\alpha_j)}x_jt_i$ and $t_iy_j = q^{-(\alpha_i,\alpha_j)}y_jt_i$ for all $1 \leq i, j \leq n$.

(1.9)

$$\sum_{m=0}^{1-a_{ij}} (-1)^m \begin{bmatrix} 1-a_{ij} \\ m \end{bmatrix}_{q_i} x_i^{1-a_{ij}-m} x_j x_i^m = 0$$

and

$$\sum_{m=0}^{1-a_{ij}} (-1)^m \begin{bmatrix} 1-a_{ij} \\ m \end{bmatrix}_{q_i} y_i^{1-a_{ij}-m} y_j y_i^m = 0$$

for all $1 \leq i, j \leq n$ with $i \neq j$.

Let U^+ denote the subalgebra of U generated by $x_i, 1 \leq i \leq n$, and let G^- denote the subalgebra of U generated by $y_i t_i, 1 \leq i \leq n$. Let U^0 denote the group algebra over $\mathcal{C}$ generated by T inside U. The triangular decomposition of U ([R]) is the following isomorphism of vector spaces induced by the multiplication map.

$$U \cong G^- \otimes U^0 \otimes U^+. \tag{1.10}$$

The algebra U is a Hopf algebra with comultiplication Δ, antipode σ, and counit ϵ. These maps are determined by the following action on the generators of U.

(1.11) $\Delta(t) = t \otimes t \qquad \epsilon(t) = 1 \qquad \sigma(t) = t^{-1}$ for all t in T

(1.12) $\Delta(x_i) = x_i \otimes 1 + t_i \otimes x_i \qquad \epsilon(x_i) = 0 \qquad \sigma(x_i) = -t_i^{-1} x_i$

(1.13) $\Delta(y_i) = y_i \otimes t_i^{-1} + 1 \otimes y_i \qquad \epsilon(y_i) = 0 \qquad \sigma(y_i) = -y_i t_i$

for $1 \leq i \leq n$.

Using Sweedler notation for the comultiplication map, we write

$$\Delta(a) = \sum a_{(1)} \otimes a_{(2)}$$

for all $a \in U$. Since U is a Hopf algebra, it admits both a left adjoint action, denoted by ad, and a right adjoint action, denoted by ad_r. In particular, given a and b in U, we have

$$(\mathrm{ad}_r\ a)b = \sum \sigma(a_{(1)}) b a_{(2)} \quad \text{and} \quad (\mathrm{ad}\ a)b = \sum a_{(1)} b \sigma(a_{(2)}).$$

The left adjoint action makes U into a left U module and the right adjoint action makes U into a right U module. Even though ad_r is a right action, we follow the conventions in [L4] and [L5], and write $(\mathrm{ad}_r\ a)$ on the left. The reader should be aware of the fact that $(\mathrm{ad}_r\ ac)b = (\mathrm{ad}_r\ c)(\mathrm{ad}_r\ a)b$ for all a, b, and c in U. It is helpful to see the description of the right adjoint action of generators of U given by:

$$\begin{aligned} (\mathrm{ad}_r\ y_i)b &= b y_i - y_i t_i b t_i^{-1} \\ (\mathrm{ad}_r\ x_i)b &= t_i^{-1} b x_i - t_i^{-1} x_i b \\ (\mathrm{ad}_r\ t)b &= t^{-1} b t \end{aligned} \tag{1.14}$$

for all $b \in U$, $t \in T$, and $1 \leq i \leq n$. Similarly, the left adjoint action satisfies the following for all $b \in U$, $t \in T$ and $1 \leq i \leq n$.

$$\begin{aligned} (\mathrm{ad}\ y_i)b &= y_i b t_i - b y_i t_i \\ (\mathrm{ad}\ x_i)b &= x_i b - t_i b t_i^{-1} x_i \\ (\mathrm{ad}\ t)b &= t b t^{-1} \end{aligned} \tag{1.15}$$

Let τ denote the group isomorphism from the additive group $Q(\pi)$ to the multiplicative group T defined by $\tau(\alpha_i) = t_i$ for $1 \leq i \leq n$. Given a vector subspace S of U, the β weight space of S denoted by S_β is the set all elements s in S which satisfy

$$\tau(\gamma) s \tau(-\gamma) = q^{(\gamma, \beta)} s \tag{1.16}$$

for all $\tau(\gamma) \in T$.

Sometimes it will be necessary to consider larger algebras than U which are obtained using extensions of the group T. In particular, suppose that M is a multiplicative monoid isomorphic to an additive submonoid of $\sum_{1\leq i\leq n}\mathbf{Q}\alpha_i$ via the obvious extension of τ. Then the algebra UM is just the algebra generated by U and M subject to the additional relation (1.16) for each $\tau(\gamma)\in M$, $\beta\in Q(\pi)$, and $s\in U_\beta$. It is straightforward to check that when M is a group, the Hopf algebra structure of U extends to UM. Moreover, the triangular decomposition (1.10) of U can be extended to UM by enlarging U^0 to U^0M. Recall that the augmentation ideal of a Hopf algebra is just the kernel of the counit. If UM is a Hopf algebra and A is a subalgebra, then we write A_+ to denote the intersection of A with the augmentation ideal of UM.

The most common extension of the type described in the previous paragraph is the simply connected quantized enveloping algebra, denoted by $\check{U}$ ([Jo, Section 3.2.10]). Here, the multiplicative monoid M is the group $\check{T}$ which is the extension of T isomorphic to the weight lattice $P(\pi)$ via τ. Let $\check{U}^0$ denote the group algebra generated by $\check{T}$.

Let $F_r(\check{U})$ denote the locally finite part of $\check{U}$ with respect to the right adjoint action ad_r. Recall that $F_r(\check{U})$ is a subalgebra of $\check{U}$ and can be written as a direct sum of all the finite-dimensional simple $(\mathrm{ad}_r\ U)$ submodules of $\check{U}$. We also have that $F_r(\check{U})\cap\check{T}$ is equal to the set $\{\tau(2\mu)|\ \mu\in P^+(\pi)\}$. Moreover, the algebra $F_r(\check{U})$ admits the following direct sum decomposition as an $(\mathrm{ad}_r\ U)$ module:

$$F_r(\check{U})=\oplus_{\lambda\in P^+(\pi)}(\mathrm{ad}_r\ U)\tau(2\lambda). \tag{1.17}$$

(For more information about the locally finite part of U, the reader is referred to [Jo, Section 7]. It should be noted that here we are using the right adjoint action while [Jo, Section 7] examines the left adjoint action. The above information has been adjusted to account for this difference.)

For each $\lambda\in\sum_{\alpha_i\in\pi}\mathbf{Q}\alpha_i$, let z^λ be the algebra homomorphism from $\check{U}^0$ to $\mathcal{C}$ defined by

$$z^\lambda(\tau(\gamma))=q^{(\lambda,\gamma)} \tag{1.18}$$

for all $\gamma\in P(\pi)$. Given $\Lambda\in\mathrm{Hom}(\check{U}^0,\mathcal{C})$, let $L(\Lambda)$ denote the simple U module with highest weight Λ. If $\Lambda=z^\lambda$ for some $\lambda\in P(\pi)$, we write $L(\Lambda)$ as $L(\lambda)$. Just as in the classical case, $L(\lambda)$ is finite dimensional for all $\lambda\in P^+(\pi)$.

For any multiplicative group G, we write $\mathcal{C}[G]$ for the group algebra generated by G over $\mathcal{C}$. This notation will be applied to groups as well as multiplicative monoids related to $\check{T}$. Now suppose that H is an additive subgroup of $\sum_{\alpha_i\in\pi}\mathbf{Q}\alpha_i$. Let $\langle z^\lambda|\ \lambda\in H\rangle$ denote the multiplicative group isomorphic to H; we write $\mathcal{C}[H]$ for its group algebra. By the previous paragraph, we may view elements of the form z^λ as functions on $\check{U}^0$. This identification allows us to interpret the group algebra $\mathcal{C}[H]$ as a subset of the dual of $\check{U}^0$.

Quantum symmetric pairs: Quantum symmetric pairs are actually analogs of the pair of enveloping algebras $U(\mathfrak{g}),U(\mathfrak{g}^\theta)$. In particular, a quantum symmetric pair consists of a pair of algebras U,B where B is a special subalgebra of U that can be viewed as a quantum analog of $U(\mathfrak{g}^\theta)$. Unfortunately, the quantized enveloping algebra U contains very few Hopf subalgebras. In general, there is no obvious way to realize $U_q(\mathfrak{g}^\theta)$ as a Hopf subalgebra of U. The philosophy in [L3] is to look for one-sided coideal subalgebras of U as quantum analogs of Hopf subalgebras of $U(\mathfrak{g})$.

Recall that a left coideal subalgebra I of U is a subalgebra of U such that

$$\Delta(I) \subseteq U \otimes I.$$

As in [L4], we call B a quantum analog of $U(\mathfrak{g}^\theta)$ inside of U provided B is a maximal left coideal subalgebra of U that specializes to $U(\mathfrak{g}^\theta)$ as q goes to 1. (See Section 6 and (6.7) for more information about specialization at $q = 1$.) Quantum analogs of $U(\mathfrak{g}^\theta)$ are classified in [L3, Section 7]; a complete list in terms of generators and relations can be found in [L4, Section 7]. We review the construction of these coideal subalgebras here. First, we describe a distinguished algebra B_θ called the standard quantum analog of $U(\mathfrak{g}^\theta)$ inside of U. For most irreducible symmetric pairs, any quantum analog of $U(\mathfrak{g}^\theta)$ inside of U is isomorphic to the standard quantum analog B_θ via a Hopf algebra automorphism of U. However, under certain circumstances which we describe below, the set of quantum analogs of $U(\mathfrak{g}^\theta)$ inside of U consists of a one parameter family of coideal subalgebras (containing B_θ) up to Hopf algebra automorphisms of U.

Define the subalgebra $\mathcal{M}$ of U as *the algebra generated by* x_i, y_i, t_i, t_i^{-1} *for* $\alpha_i \in \pi_\Theta$. Note that $\mathcal{M}$ is just the quantized enveloping algebra of the semisimple Lie subalgebra $\mathfrak{m}$ of $\mathfrak{g}$ generated by e_i, f_i, h_i with $\alpha_i \in \pi_\Theta$. In particular, $\mathcal{M}$ is a Hopf subalgebra of U. Let T_Θ be the subgroup of T defined by

$$T_\Theta = \{\tau(\mu) |\ \mu \in Q(\pi) \text{ and } \Theta(\mu) = \mu\}.$$

One checks that the group algebra $\mathbf{C}(q)[T_\Theta]$ specializes to the enveloping algebra $U(\mathfrak{h}^\theta)$ where $\mathfrak{h}^\theta = \mathfrak{h} \cap \mathfrak{g}^\theta$. In the classical case, one can show that $U(\mathfrak{g}^\theta)$ is generated as an algebra by $\mathfrak{m}, \mathfrak{h}^\theta$, and the elements $f_i + \theta(f_i)$ for $\alpha_i \in \pi \setminus \pi_\Theta$. The standard quantum analog of $U(\mathfrak{g}^\theta)$ inside of U is generated by $\mathcal{M}$, T_Θ, and lifts of the elements $f_i + \theta(f_i)$, for $\alpha_i \in \pi \setminus \pi_\Theta$. These "lifts" are constructed by first lifting the involution θ on $\mathfrak{g}$ to a $\mathbf{C}$ algebra automorphism $\tilde{\theta}$ of U sending q to q^{-1} ([L3, Theorem 7.1]). The action of $\tilde{\theta}$ on the y_i for $\alpha_i \in \pi \setminus \pi_\Theta$ can be described explicitly as follows. Recall the notion of divided powers of the x_i and y_i defined by

$$x_i^{(m)} = \frac{x_i^m}{[m]!_{q_i}} \text{ and } y_i^{(m)} = \frac{y_i^m}{[m]!_{q_i}}$$

for each $1 \leq i \leq n$. Given $\alpha_i \in \pi^*$, we can find a sequence $\alpha_{i_1}, \ldots, \alpha_{i_s}$ of roots in π_Θ and positive integers $m_1, \ldots, m_s$ such that

$$\tilde{\theta}(y_i) = (\mathrm{ad}_r\ x_{i_1}^{(m_1)}) \cdots (\mathrm{ad}_r\ x_{i_s}^{(m_s)}) t_{\mathrm{p}(i)}^{-1} x_{\mathrm{p}(i)} \tag{1.19}$$

and

$$\tilde{\theta}(y_{\mathrm{p}(i)}) = (-1)^{(m_1 + \cdots + m_s)} (\mathrm{ad}_r\ x_{i_s}^{(m_s)}) \cdots (\mathrm{ad}_r\ x_{i_1}^{(m_1)}) t_i^{-1} x_i. \tag{1.20}$$

Note that in this notation, the evaluation of Θ on α_i is given by

$$\Theta(\alpha_i) = -m_1 \alpha_{i_1} \cdots - m_s \alpha_{i_s} - \alpha_{\mathrm{p}(i)}.$$

In the special case when $\Theta(\alpha_i) = -\alpha_i$, we see that $\tilde{\theta}(y_i) = t_i^{-1} x_i$. (More generally, the specific sequences of simple roots in π_Θ and positive integers m_i associated to a particular irreducible symmetric pair can be read off the information in [L4, Section 7]. See also the discussion in [L3, Section 7].)

Set

$$B_i = y_i t_i + \tilde{\theta}(y_i) t_i$$

for all $\alpha_i \in \pi \setminus \pi_\Theta$. The standard quantum analog of $U(\mathfrak{g}^\theta)$ inside of U is the subalgebra B_θ of U defined to be *the algebra generated by $\mathcal{M}, T_\Theta$, and B_i for $\alpha_i \in \pi \setminus \pi_\Theta$*. The relations satisfied by the generators of B_θ can be deduced directly from the relations of U, though some of the computations are quite lengthy. The relations satisfied by the B_i, for $\alpha_i \notin \pi_\Theta$, and the other generators of B_θ are given in [L4, Section 7].

Define two subsets $\mathcal{S}$ and $\mathcal{D}$ of π^* by

$$\mathcal{S} = \{\alpha_i \in \pi^* |\ \Theta(\alpha_i) = -\alpha_i, \text{ and } \frac{(\alpha_i, \alpha_j)}{(\alpha_j, \alpha_j)} \in \mathbf{Z} \text{ whenever } \Theta(\alpha_j) = -\alpha_j\}$$

and

$$\mathcal{D} = \{\alpha_i \in \pi^* |\ i \neq \mathrm{p}(i) \text{ and } (\alpha_i, \Theta(\alpha_i)) \neq 0\}.$$

Since we are assuming that $\mathfrak{g}, \mathfrak{g}^\theta$ is irreducible, it follows that $\mathcal{S} \cup \mathcal{D}$ has at most one element ([L4, Section 2 and Section 7]). Suppose first that $\mathcal{S}$ is not empty. In particular, $\mathcal{S}$ contains a single root, say α_j. By the definition of $\mathcal{S}$, we have that $\Theta(\alpha_j) = -\alpha_j$. It follows from the discussion above concerning $\tilde{\theta}$ that $\tilde{\theta}(y_j) = t_j^{-1} x_j$. Fix $s \in \mathcal{C}$ and set

$$\begin{aligned} B_{j,s,1} &= y_j t_j + \tilde{\theta}(y_j) t_j + s t_j \\ &= y_j t_j + t_j^{-1} x_j t_j + s t_j. \end{aligned}$$

Let $B_{\theta,s,1}$ be *the subalgebra of U generated by $\mathcal{M}, T_\Theta, B_i$ for α_i in the set $\pi \setminus (\pi_\Theta \cup \mathcal{S})$ and $B_{j,s,1}$*. By ([L3, Section 7, Variation 2]), $B_{\theta,s,1}$ is a left coideal subalgebra of U for all $s \in \mathcal{C}$. Note that when $s = 0$, the algebra $B_{\theta,0,1}$ is just the standard quantum analog B_θ. More generally, if s goes to 0 as q goes to 1, then $B_{\theta,s,1}$ is a quantum analog of $U(\mathfrak{g}^\theta)$ inside of U.

Now consider the case when $\mathcal{D}$ is nonempty and write $\mathcal{D} = \{\alpha_j\}$. Fix $d \in \mathcal{C}$ such that $d \neq 0$. Set

$$B_{j,0,d} = y_j t_j + d\tilde{\theta}(y_j) t_j.$$

Let $B_{\theta,0,d}$ be *the subalgebra of U generated by $\mathcal{M}, T_\Theta, B_i$ for α_i in the set $\pi \setminus (\pi_\Theta \cup \mathcal{D})$, and $B_{j,0,d}$*. By ([L3, Section 7, Variation 1]), $B_{\theta,0,d}$ is a left coideal subalgebra of U. In the special case when $d = 1$, we have that $B_{\theta,0,1}$ is just the standard analog B_θ. More generally, any quantum analog of $U(\mathfrak{g}^\theta)$ is isomorphic via a Hopf algebra automorphism of U to $B_{\theta,0,d}$ for some scalar d that goes to 1 as q goes to 1.

Let $U_\mathcal{R}$ be the $\mathcal{R}$ subalgebra of U generated by $x_j, y_j, t_j^{\pm 1}$ for $1 \leq j \leq n$. A subalgebra B of U is called *real* if $B = (B \cap U_\mathcal{R}) \oplus i(B \cap U_\mathcal{R})$ (where here i denotes the square root of -1). Note that the standard analog B_θ is real since it is generated by elements of $U_\mathcal{R}$. More generally, when $\mathcal{S}$ is nonempty, $B_{\theta,s,1}$ is real provided $s \in \mathcal{R}$. Similarly, when $\mathcal{D}$ is nonempty, $B_{\theta,0,d}$ is real provided $d \in \mathcal{R}$.

Set $\mathcal{S}_\mathcal{R}$ equal to $\mathcal{R}$ if $\mathcal{S}$ is nonempty and equal to $\{0\}$ otherwise. Similarly, set $\mathcal{D}_\mathcal{R}$ equal to $\mathcal{R} \setminus \{0\}$ if $\mathcal{D}$ is nonempty and equal to $\{1\}$ otherwise. Consider the set of coideal subalgebras

$$\{B_{\theta,s,d} | s \in \mathcal{S}_\mathcal{R}, d \in \mathcal{D}_\mathcal{R}\}. \tag{1.21}$$

In the case when $\mathcal{S} \cup \mathcal{D}$ is empty, the set in (1.21) consists of the single algebra B_θ. Otherwise, the set (1.21) is a one parameter family of coideal subalgebras depending on s if $\mathcal{S}$ is nonempty and on d if $\mathcal{D}$ is nonempty. By [L3, Theorem 7.5], any real quantum analog of $U(\mathfrak{g}^\theta)$ inside U is isomorphic via a Hopf algebra automorphism of U to some element of the set (1.21). It should be noted, however, that not every element in this set is a quantum analog of $U(\mathfrak{g}^\theta)$ inside U since not all the subalgebras in $\mathcal{B}$ specialize to $U(\mathfrak{g}^\theta)$ as q goes to 1. For example, if $\mathcal{S}$ is nonempty, then $B_{\theta,s,1}$ specializes to $U(\mathfrak{g}^\theta)$ if and only if s specializes to 0.

Let $\mathbf{H}$ denote the set of Hopf algebra automorphisms of U that fix elements of T and restrict to an automorphism of $U_{\mathcal{R}}$. Note that $\mathbf{H}$ acts on the set of coideal subalgebras of U. Set $\mathcal{B}$ equal to the orbit of the set (1.21) under $\mathbf{H}$. As explained in the previous paragraph, we see that $\mathcal{B}$ contains all real quantum analogs of $U(\mathfrak{g}^\theta)$ inside of U. However, $\mathcal{B}$ also contains coideal subalgebras that do not specialize to $U(\mathfrak{g}^\theta)$ as q goes to 1. Despite this fact, we refer to the set $\mathcal{B}$ as the set of real quantum analogs of $U(\mathfrak{g}^\theta)$ inside of $U(\mathfrak{g})$.

Given $\alpha_i \in \mathcal{S}$, we shorten $B_{i,s,1}$ to B_i when s can be read from context. A similar convention is applied to $B_{i,0,d}$ for $\alpha_i \in \mathcal{D}$. Set $s_i = s$ for $\alpha_i \in \mathcal{S}$ and $s_i = 0$ for all $\alpha_i \notin \mathcal{S}$. Similarly, set $d_i = d$ for $\alpha_i \in \mathcal{D}$ and set $d_i = 1$ for all $\alpha_i \notin \mathcal{D}$. In order to make arguments in the general case, it will often be convenient to write each B_i as

$$B_i = y_i t_i + d_i \tilde{\theta}(y_i) t_i + s_i t_i.$$

The $*$ structure: Fix a left coideal subalgebra B in $\mathcal{B}$. An important tool in analyzing B is a conjugate linear antiautomorphism of $\check{U}$ which restricts to a conjugate linear antiautomorphism of B. It is constructed in [L4, Theorem 3.1] from a quantum version of the Chevalley antiautomorphism of $\mathfrak{g}$. More precisely, let κ_1 denote the antiautomorphism of $U_{\mathcal{R}}$ defined by

$$\kappa_1(x_j) = y_j t_j \quad \kappa_1(y_j) = t_j^{-1} x_j \quad \kappa_1(t) = t$$

for all $t \in T$ and $1 \leq j \leq n$. Using the decomposition $U = U_{\mathcal{R}} \oplus iU_{\mathcal{R}}$, we can extend κ_1 to a conjugate linear antiautomorphism of U. Theorem 3.1 of [L4] shows that there exists a Hopf algebra automorphism ψ in $\mathbf{H}$ such that $\psi\kappa_1\psi^{-1}$ restricts to a conjugate linear antiautomorphism of B. We set $\kappa_B = \psi\kappa_1\psi^{-1}$ and write κ for κ_B when B can be understood from context. Note that there exist nonzero scalars $c_{Bj} \in \mathcal{R}$ for $1 \leq j \leq n$ such that

$$\kappa_B(x_j) = c_{Bj} y_j t_j \quad \kappa_B(y_j) = c_{Bj}^{-1} t_j^{-1} x_j \quad \kappa_B(t) = t \tag{1.22}$$

for all $t \in T$ and $1 \leq j \leq n$.

As explained in [L3, Section 2] (see also [CP, Section 4.1F]), κ_B gives U the structure of a Hopf $*$ algebra where we take $*$ equal to κ_B. Since $\kappa_B(B) = B$, it follows that B is a $*$ subalgebra of U. By [L3, Theorem 2.4], we know that B acts semisimply on finite-dimensional U modules. Given a finite-dimensional U module V, we write V^B for the subspace of B invariant elements. In particular,

$$V^B = \{v \in V \mid bv = \epsilon(b)v \text{ for all } b \in B\} = \{v \in V \mid B_+ v = 0\}.$$

It follows that each $v \in V^B$ generates a one-dimensional trivial B module. Thus V^B can be expressed as a direct sum of one-dimensional trivial B modules. The fact that B acts semisimply on V yields the following direct sum decomposition of V into B modules.

$$V = V^B \oplus B_+ V. \tag{1.23}$$

The U module V is called *spherical* provided V^B is nonzero. Finite-dimensional simple spherical U modules are classified in [L4, Theorem 3.2] as follows.

THEOREM 1.1. *Let $\lambda \in P^+(\pi)$. Then for all $B \in \mathcal{B}$, we have* $\dim L(\lambda)^B \leq 1$. *Moreover,* $\dim L(\lambda)^B = 1$ *if and only if $\lambda = 2\mu$ for some $\mu \in P^+(\Sigma)$.*

Note that the map κ_1, as well as the map κ_B, can be extended to a conjugate linear antiautomorphism of $\check{U}$ such that $\kappa_B(t) = \kappa_1(t) = t$ for all $t \in \check{T}$. Since the locally finite part $F_r(\check{U})$ of $\check{U}$ is semisimple, it follows that B acts semisimply on $F_r(\check{U})$. Moreover, the map κ_B is used to prove the following theorem which is a combination of [L3, Theorem 7.6] and [L2, Theorem 3.5].

THEOREM 1.2. *For all $B \in \mathcal{B}$, $F_r(\check{U})$ can be written as a direct sum of finite-dimensional simple* $(\mathrm{ad}_r\ B)$ *modules. Moreover, if V is a finite-dimensional* $(\mathrm{ad}_r\ B)$ *submodule of $\check{U}$ then V is a subset of $F_r(\check{U})$.*

The antiautomorphism κ_B behaves particularly well with respect to the right adjoint action of U. Using (1.14) and (1.22), we see that

$$\begin{aligned}\kappa_1((\mathrm{ad}_r\ y_i)u) &= \kappa_1(uy_i - y_it_iut_i^{-1})\\ &= t_i^{-1}x_i\kappa_1(u) - t_i^{-1}\kappa_1(u)x_i\\ &= -(\mathrm{ad}_r\ x_i)\kappa_1(u)\end{aligned}$$

for all $u \in \check{U}$ and for all $1 \leq i \leq n$. Similar computations yields $\kappa_1((\mathrm{ad}_r\ x_i)u) = -(\mathrm{ad}_r\ y_i)\kappa_1(u)$ and $\kappa_1((\mathrm{ad}_r\ t)u) = (\mathrm{ad}_r\ t^{-1})\kappa_1(u)$ for all $t \in T$. Following the notation of (1.22), we see that

$$\begin{aligned}\kappa_B((\mathrm{ad}_r\ x_i)u) &= -\ c_{Bi}(\mathrm{ad}_r\ y_i)\kappa_B(u)\\ \kappa_B((\mathrm{ad}_r\ y_i)u) &= -\ c_{Bi}^{-1}(\mathrm{ad}_r\ x_i)\kappa_B(u)\\ \kappa_B(tut^{-1}) &= (\mathrm{ad}_r\ t^{-1})\kappa_B(u)\end{aligned} \tag{1.24}$$

for each $u \in \check{U}$, $t \in T$, and $1 \leq i \leq n$. By (1.24) we see that the image of an $\mathrm{ad}_r\ U$ module under κ_B is also an $\mathrm{ad}_r\ U$ module. In particular, we have

$$\kappa_B(F_r(\check{U})) = F_r(\check{U}) \tag{1.25}$$

for all $B \in \mathcal{B}$.

Quantum (compact) symmetric spaces: Recall that there is no quantum version of the connected, simply connected algebraic Lie group with Lie algebra $\mathfrak{g}$. Instead we study a quantum analog of the ring of regular functions on this Lie group. Similarly, there is no quantum version of the compact symmetric space defined using the (compact) Riemannian symmetric pair associated to $\mathfrak{g}, \mathfrak{g}^\theta$. Instead, we consider quantum analogs of the ring of regular functions on this symmetric space. More precisely, let $R_q[G]$ denote the quantized function algebra associated to $\mathfrak{g}$ (see [Jo, Section 9.1] for exact definition). Consider a subalgebra $B \in \mathcal{B}$. The quantum analog of the ring of regular functions on the compact symmetric space associated to $\mathfrak{g}, \mathfrak{g}^\theta$ is the set $R_q[G/K]_B$ of right B invariants inside $R_q[G]$. Note that $R_q[G/K]_B$ is not a Hopf subalgebra of the quantized function algebra. However, it is a right coideal subalgebra of $R_q[G]$ ([L3, Section 3]).

Let $Z(\check{U})$ denote the center of $\check{U}$. The quantum (compact) zonal spherical functions defined by a pair B, B' in $\mathcal{B}$ are nonzero left B' and right B invariant elements of $R_q[G]$ that are also simultaneous eigenfunctions with respect to the action of $Z(\check{U})$. Since the quantum zonal spherical functions are right B invariant, they can also be viewed as elements of $R_q[G/K]_B$. We give a more explicit description of these spherical functions below using the quantum Peter-Weyl theorem.

Given $\lambda \in P^+(\pi)$, let $L(\lambda)^*$ denote the dual of the finite-dimensional simple U module $L(\lambda)$. The vector space $L(\lambda)^*$ is a U module given the natural right U module structure. The quantum Peter-Weyl theorem [Jo, 9.1.1 and 1.4.13] (see also [L3, (3.1)]) is the following isomorphism of U bimodules:

$$R_q[G] \cong \bigoplus_{\lambda \in P^+(\pi)} L(\lambda) \otimes L(\lambda)^*. \tag{1.26}$$

Note that a central element acts as a scalar on both the right and left on $L(\lambda) \otimes L(\lambda)^*$. Moreover, this scalar is determined by the central character associated to the finite-dimensional simple module $L(\lambda)$. Thus each $L(\lambda) \otimes L(\lambda)^*$ is a joint eigenspace for the action of the center on $R_q[G]$. Identifying $R_q[G]$ with the right hand side of (1.26) allows us to interpret elements of $R_q[G]$ as functions on $\check{U}$. In particular, $(v \otimes v^*)(u) = v^*(uv)$ for all $v \in L(\lambda), v^* \in L(\lambda)^*, u \in \check{U}$, and $\lambda \in P^+(\pi)$.

Fix B in $\mathcal{B}$ and assume that B' is another subalgebra of $\mathcal{B}$. Let ${}_{B'}\mathcal{H}_B$ denote the subspace of $R_q[G]$ consisting of left B' and right B invariants. Set

$${}_{B'}\mathcal{H}_B(\lambda) = {}_{B'}\mathcal{H}_B \cap (L(\lambda) \otimes L(\lambda)^*).$$

Theorem 1.1 implies that ${}_{B'}\mathcal{H}_B(\mu)$ is one dimensional if $\mu = 2\lambda$ for some $\lambda \in P^+(\Sigma)$ and equals zero otherwise. Hence, the Peter-Weyl theorem (1.26) allows us to decompose ${}_{B'}\mathcal{H}_B$ into a direct sum of left B' and right B modules

$${}_{B'}\mathcal{H}_B = \bigoplus_{\lambda \in P^+(\Sigma)} {}_{B'}\mathcal{H}_B(2\lambda).$$

Note that this is also a decomposition of ${}_{B'}\mathcal{H}_B$ into joint eigenspaces with respect to the action of the center $Z(\check{U})$ on $R_q[G]$.

As explained above, we can view elements of ${}_{B'}\mathcal{H}_B$ as functions on the simply connected quantized enveloping algebra $\check{U}$. Restriction to $\check{U}^0$ yields a map Υ from $R_q[G]$ into the dual vector space of $\check{U}^0$. Let $\mathcal{C}[P(2\Sigma)]$ denote the group algebra $\mathcal{C}[z^{2\mu} |\ \mu \in P(\Sigma)]$. Using (1.18) we can identify $\mathcal{C}[P(2\Sigma)]$ with a subspace of the dual of $\check{U}^0$. By [L4, Theorem 4.2], the restriction of Υ to ${}_{B'}\mathcal{H}_B$ is an algebra homomorphism

$$\Upsilon : {}_{B'}\mathcal{H}_B \mapsto \mathcal{C}[P(2\Sigma)].$$

Let W_Θ denote the Weyl group of the restricted root system Σ. The action of W_Θ on Σ induces an action of W_Θ on $\mathcal{C}[P(2\Sigma)]$ such that $w(z^{2\mu}) = z^{2w\mu}$ for all $w \in W_\Theta$. Using [L4, Theorem 5.3 and Corollary 5.4], we may choose B' so that the image of ${}_{B'}\mathcal{H}_B$ under Υ is W_Θ invariant. In particular, let ρ denote the half sum of the positive roots in Δ. Let χ denote the Hopf algebra automorphism of U defined by

$$\chi(x_i) = q^{(\rho,\tilde{\alpha}_i)}x_i, \quad \chi(y_i) = q^{-(\rho,\tilde{\alpha}_i)}y_i, \quad \text{and} \quad \chi(t) = t \tag{1.27}$$

for all $1 \le i \le n$ and $t \in T$. The following is [L4, Corollary 5.4].

THEOREM 1.3. *For all $B \in \mathcal{B}$, the restriction of Υ to ${}_{\chi(B)}\mathcal{H}_B$ is an algebra isomorphism of ${}_{\chi(B)}\mathcal{H}_B$ onto the ring $\mathcal{C}[P(2\Sigma)]^{W_\Theta}$.*

Consider for the moment the special case when $\mathcal{S} \cup \mathcal{D}$ is nonempty. In particular, there is a one parameter family of quantum analogs of $U(\mathfrak{g}^\theta)$ inside of U (see (1.21)). When this happens, one can find a one parameter family of algebras $\{B_t\}$

(depending on the original fixed subalgebra B) in $\mathcal{B}$ such that the image of each ${}_{B_t}\mathcal{H}_B$ under Υ is W_Θ invariant.

Given $\lambda \in P^+(\Sigma)$, let $g^{2\lambda}_{B,B'}$ denote a basis vector for ${}_{B'}\mathcal{H}_B(2\lambda)$. The element $g^{2\lambda}_{B,B'}$ (and more generally, any nonzero multiple of $g^{2\lambda}_{B,B'}$) is called a quantum (compact) zonal spherical function at 2λ associated to the pair B, B'. We will mainly be interested in the special case when $B' = \chi(B)$. For each $\lambda \in P^+(\Sigma)$, we denote $g^{2\lambda}_{B,\chi(B)}$ by $g_{2\lambda}$. Of course, this definition of $g_{2\lambda}$ depends on the choice of B, but this choice will be understood from context. Set $\varphi_{2\lambda} = \Upsilon(g_{2\lambda})$ for each λ. It should be noted that for most symmetric pairs and up to multiplication by a nonzero constant, $\varphi_{2\lambda}$ is independent of the choice of B ([L4, Theorem 6.5 and subsequent discussion]). Theorem 1.3 ensures that $\{\varphi_{2\lambda} |\ \lambda \in P^+(\Sigma)\}$ is a basis for $\mathcal{C}[P(2\Sigma)]^{W_\Theta}$. This special basis of $\mathcal{C}[P(2\Sigma)]^{W_\Theta}$ can be identified with a family of Macdonald polynomials ([L5, Theorem 8.2]) when the restricted root system is reduced. Preliminary work shows that a similar result holds when Σ is not reduced. When $\mathcal{S} \cup \mathcal{D}$ is nonempty, the set $\{\varphi_{2\lambda} |\ \lambda \in P^+(\Sigma)\}$ is part of a larger family of W_Θ invariant polynomials associated to the spaces ${}_{B_t}\mathcal{H}_B$ as t varies over $\mathcal{R}$ and B varies over elements in the set (1.21) (see [L4, Section 6]). We expect this larger family of W_Θ invariant elements to be identified with a family of Macdonald-Koornwinder polynomials associated to root systems of type BC. Indeed this follows in most cases from [DN], [NDS], and [DS].

Recall that the classical invariant differential operators correspond to ad $\mathfrak{g}^\theta$ invariant elements of $U(\mathfrak{g})$. Fix $B \in \mathcal{B}$. Let $\check{U}^B$ denote the space of B invariants inside of $\check{U}$ with respect to the right adjoint action. We view elements of $\check{U}^B$ as quantum invariant differential operators associated to the quantum symmetric pair U, B.

Note that $\check{U}^B$ is just the set of all one-dimensional trivial $(\mathrm{ad}_r\ B)$ modules. By Theorem 1.2, $\check{U}^B$ is a subset of $F_r(\check{U})$. We further have ([L5, Lemma 3.5])

$$\check{U}^B = \{u \in \check{U} |\ ub = bu \text{ for all } b \in B\} = \mathrm{Cent}_B(\check{U}). \tag{1.28}$$

It follows that $\check{U}^B$ is a subalgebra of $\check{U}$. Moreover, $Z(\check{U})$ is a subalgbra of $\check{U}^B$.

In the classical case, the zonal spherical functions are joint eigenfuctions with respect to the action of the invariant differential operators. The analogous result is true in the quantum case. (The proof for this can also be found in the discussion preceding [L5, Theorem 3.6].)

THEOREM 1.4. *The zonal spherical functions* $\{g_{2\lambda} |\ \lambda \in P^+(\Sigma)\}$ *are joint eigenfunctions with respect to the right action of* $\check{U}^B$ *on* ${}_{\chi(B)}\mathcal{H}_B$.

PROOF. Suppose that V is a right U module and v is a B invariant vector inside of V. It follows that vc is also B invariant for each $c \in \check{U}^B$. Now suppose that $\lambda \in P^+(\Sigma)$. By Theorem 1.1, $(L(2\lambda)^*)^B$ is one-dimensional. Hence elements of $\check{U}^B$ act as scalars on $(L(2\lambda)^*)^B$. The result now follows from the fact that the zonal spherical function $g_{2\lambda}$ is contained in $L(2\lambda) \otimes (L(2\lambda)^*)^B$. $\square$

CHAPTER 2

A Comparison of Two Root Systems

This section analyzes the image of the weight lattice $P(\pi)$ under the map $\tilde{}$ introduced in (1.4) and compares this image with the restricted weight lattice $P(\Sigma)$. The image of the fundamental weights associated to π under the restriction map $\tilde{}$ are completely determined. As a consequence, we establish necessary and sufficient conditions for $\widetilde{P^+(\pi)}$ to equal $P^+(\Sigma)$. This is an important first step in understanding the image of the center $Z(\check{U})$ under the Harish-Chandra map associated to a quantum symmetric pair. We also provide detailed descriptions of the restricted roots and corresponding restricted fundamental weights associated to those symmetric pairs $\mathfrak{g}, \mathfrak{g}^\theta$ with $\widetilde{P^+(\pi)}$ not equal to $P^+(\Sigma)$. This information is used in the analysis of the quantum Harish-Chandra map for certain families of symmetric pairs in Sections 8 and 9.

For each $1 \leq i \leq n$, set ω_i equal to the fundamental weight corresponding to the simple root α_i. Recall that the restricted root system is either an ordinary reduced root system as classified in [H, Chapter III] or is nonreduced of type BC_r for some integer $r \geq 1$ ([Kn, Chapter II, Section 5]). In the latter case, there exist roots $\alpha \in \Delta$ such that both $\tilde{\alpha}$ and $2\tilde{\alpha}$ are restricted roots in Σ. Given $\alpha_i \in \pi^*$, let ω_i' denote the fundamental weight corresponding to the simple root $\tilde{\alpha}_i$ with respect to the restricted root system. In particular, if $2\tilde{\alpha}_i$ is not a restricted root then $(\omega_i', \tilde{\alpha}_i) = (\tilde{\alpha}_i, \tilde{\alpha}_i)/2$ and if $2\tilde{\alpha}_i$ is a restricted root then $(\omega_i', 2\tilde{\alpha}_i) = (2\tilde{\alpha}_i, 2\tilde{\alpha}_i)/2$.

We can break up the set $\{\tilde{\alpha}_i | \ \alpha_i \in \pi^*\}$ into three cases. In each case, we give the value of $(\tilde{\alpha}_i, \tilde{\alpha}_i)$.

Case 1: $\Theta(\alpha_i) = -\alpha_i$. Then $(\tilde{\alpha}_i, \tilde{\alpha}_i) = (\alpha_i, \alpha_i)$.

Case 2: $\Theta(\alpha_i) \neq -\alpha_i$, and $(\alpha_i, \Theta(\alpha_i)) = 0$. Then $\tilde{\alpha}_i = (\alpha_i - \Theta(\alpha_i))/2$ and $(\tilde{\alpha}_i, \tilde{\alpha}_i) = (\alpha_i, \alpha_i)/2$.

Case 3: $\Theta(\alpha_i) \neq -\alpha_i$, and $(\alpha_i, \Theta(\alpha_i)) \neq 0$. In this case, $\alpha_i + \Theta(-\alpha_i)$ is a root. Note that $\tilde{\alpha}_i = \widetilde{\Theta(-\alpha_i)}$. Hence, both $\tilde{\alpha}_i$ and $2\tilde{\alpha}_i$ are roots in Σ. In particular, as an abstract root system, Σ must be of type BC_r for some r. The only way this can happen is for $(\alpha_i, \Theta(-\alpha_i)) = -(\alpha_i, \alpha_i)/2$. It follows that $(\tilde{\alpha}_i, \tilde{\alpha}_i) = (\alpha_i, \alpha_i)/4$.

LEMMA 2.1. *Suppose $\alpha_i \in \pi^*$. If $i = \mathrm{p}(i)$ and $\alpha_i \neq \Theta(-\alpha_i)$ then $\tilde{\omega}_i = 2\omega_i'$. If $i \neq \mathrm{p}(i)$ or if $\alpha_i = \Theta(-\alpha_i)$ then $\tilde{\omega}_i = \omega_i'$.*

PROOF. Fix $\alpha_i \in \pi^*$. Note that $(\tilde{\omega}_i, \tilde{\alpha}_j) = (\omega_i, \tilde{\alpha}_j) = 0$ whenever $\tilde{\alpha}_j \neq \tilde{\alpha}_i$. It follows that $\tilde{\omega}_i$ is a scalar multiple of ω_i'. Now

$$(\tilde{\omega}_i, \tilde{\alpha}_i) = (\omega_i, \tilde{\alpha}_i) = (1 + \delta_{i\mathrm{p}(i)})(\alpha_i, \alpha_i)/4.$$

Note that we can break up both Case 2 and Case 3 into two subcases depending on whether $i = \mathrm{p}(i)$ or $i \neq \mathrm{p}(i)$. The lemma follows from a straightforward computation using the three cases above and the corresponding subcases. □

Let W denote the Weyl group associated to the root system of $\mathfrak{g}$ and recall that W_Θ is the Weyl group associated to the restricted root system Σ. It is well known that the restricted Weyl group W_Θ is a homomorphic image of the subgroup of W which leaves $Q(\Sigma)$ invariant (see [He, p. 791]). In particular, given $\tilde{w}$ in W_Θ, there exists w in W such that $\tilde{w}$ and w act the same on $Q(\Sigma)$.

Note that $P(\pi)$ is generated by ω_i for $1 \le i \le n$. Similarly, $P(\Sigma)$ is generated by ω_i' for $\alpha_i \in \pi^*$. It follows from Lemma 2.1 that the sublattice of $\widetilde{P(\pi)}$ generated by $\tilde{\omega}_i$ for $\alpha_i \in \pi^*$ is contained in $P(\Sigma)$. The next lemma extends this result to all of $\widetilde{P(\pi)}$.

LEMMA 2.2. *The set $\widetilde{P(\pi)}$ is a subset of $P(\Sigma)$.*

PROOF. Let $\omega = \sum_j m_j\omega_j$ be an element of $P(\pi)$. We need to show that $(\tilde{\omega}, \tilde{\alpha}) \in \mathbf{Z}(\tilde{\alpha}, \tilde{\alpha})/2$ for all $\tilde{\alpha} \in \Sigma$. Now suppose $\tilde{\alpha} \in \Sigma$. We can find $\tilde{w} \in W_\Theta$ such that $\tilde{w}\tilde{\alpha} = \tilde{\alpha}_i$ or $\tilde{w}\tilde{\alpha} = 2\tilde{\alpha}_i$ for some $\alpha_i \in \pi^*$. (The latter case can occur only if Σ is of type BC_r for some r and $\tilde{\alpha}/2$ is also a root.) By the discussion preceding the lemma, there exists $w \in W$ such that w agrees with $\tilde{w}$ upon restriction to $Q(\Sigma)$. Since $P(\pi)$ is invariant under the action of W, we may assume that $\tilde{\alpha}$ is either $\tilde{\alpha}_i$ or $2\tilde{\alpha}_i$ for some $\alpha_i \in \pi^*$.

By (1.5), we can write

$$\tilde{\alpha}_i = (\alpha_i + \alpha_{\mathrm{p}(i)} + \sum_{\alpha_j \in \pi_\Theta} r_j\alpha_j)/2$$

for some nonnegative integers r_j. Note that $(\tilde{\gamma}, \tilde{\beta}) = (\gamma, \tilde{\beta})$ for any weights γ and β. It follows that

$$(\tilde{\omega}, \tilde{\alpha}_i) = (m_i r_i(\alpha_i, \alpha_i) + m_{\mathrm{p}(i)} r_{\mathrm{p}(i)}(\alpha_{\mathrm{p}(i)}, \alpha_{\mathrm{p}(i)}) + \sum_{\alpha_j \in \pi_\Theta} m_j r_j(\alpha_j, \alpha_j))/4.$$

Note that if α_i satisfies the conditions of Case 1, then $r_j = 0$ for all $\alpha_j \in \pi_\Theta$ and $\alpha_i = \alpha_{\mathrm{p}(i)}$. Hence $(\tilde{\omega}, \tilde{\alpha}_i) = m_i r_i(\alpha_i, \alpha_i)/2 \in \mathbf{Z}(\tilde{\alpha}_i, \tilde{\alpha}_i)/2$. Now assume α_i satisfies the conditions of Case 2 or Case 3. From the list in Araki's paper [A], one checks that $\mathfrak{g}$ is not of type G_2. Hence for $\alpha_j \in \pi_\Theta$, $(\alpha_j, \alpha_j) = s(\alpha_i, \alpha_i)$ where $s = 1, 2,$ or $1/2$. Moreover, this last possibility occurs if and only if $\mathfrak{g}$ is of type B_r and $\Theta(\alpha_i) = -(\alpha_i + 2\alpha_{i+1} + \cdots + 2\alpha_r)$. It follows that $\sum_{\alpha_j \in \pi_\Theta} m_j r_j(\alpha_j, \alpha_j) \in \mathbf{Z}(\alpha_i, \alpha_i)$. Since p induces a diagram automorphism on π, we have that $(\alpha_{\mathrm{p}(i)}, \alpha_{\mathrm{p}(i)}) = (\alpha_i, \alpha_i)$. Thus

$$(\tilde{\omega}, \tilde{\alpha}_i) \in \mathbf{Z}(\alpha_i, \alpha_i)/4. \tag{2.1}$$

It follows that $(\tilde{\omega}, \tilde{\alpha}_i) \in \mathbf{Z}(\tilde{\alpha}_i, \tilde{\alpha}_i)/2$ for all α_i which satisfies the conditions of Case 2. Now assume that α_i satisfies the conditions of Case 3 and so $2\tilde{\alpha}_i$ is also a positive restricted root. In this case, $(\tilde{\alpha}_i, \tilde{\alpha}_i) = (\alpha_i, \alpha_i)/4$. Therefore (2.1) implies that $(\tilde{\omega}, \tilde{\alpha}_i) \in \mathbf{Z}(\tilde{\alpha}_i, \tilde{\alpha}_i)$ and $(\tilde{\omega}, 2\tilde{\alpha}_i) \in \mathbf{Z}(2\tilde{\alpha}_i, 2\tilde{\alpha}_i)/2$. □

An immediate consequence of Lemma 2.2 is that $\widetilde{P^+(\pi)}$ is a subset of $P^+(\Sigma)$. The next result will be used to determine which irreducible symmetric pairs satisfy $\widetilde{P^+(\pi)} = P^+(\Sigma)$.

LEMMA 2.3. *Assume $\mathfrak{g}, \mathfrak{g}^\theta$ is not of type FII. Suppose that each $\alpha_i \in \pi^*$ satisfies one of the following conditions.*

(i) $\Theta(\alpha_i) = -\alpha_i$

(ii) $i \neq \mathrm{p}(i)$
(iii) $\Theta(\alpha_j) = -\alpha_{\mathrm{p}(j)}$ *for all* $j \neq i$

Then $\widetilde{P^+(\pi)} = P^+(\Sigma)$.

PROOF. Fix i so that $\alpha_i \in \pi^*$. If α_i satisfies (i) or (ii) then Lemma 2.1 implies that $\tilde{\omega}_i = \omega_i'$. Now assume that $i = \mathrm{p}(i)$ and $\Theta(\alpha_i) \neq -\alpha_i$. Assume further that α_i satisfies the conditions of (iii). Then $\pi_\Theta \cap \{\alpha_k | \ (\omega_k, \tilde{\alpha}_j) \neq 0\}$ is the empty set for $j \neq i$. Hence $(\omega_k, \tilde{\alpha}_j) = 0$ for $\alpha_k \in \pi_\Theta$ and $j \neq i$. Thus, it is sufficient to show that $(\omega_i', \tilde{\alpha}_i) = (\tilde{\lambda}, \tilde{\alpha}_i)$ for some $\lambda \in \{\omega_k | \alpha_k \in \pi_\Theta\}$. In particular, without loss of generality, we may assume that the symmetric pair $\mathfrak{g}, \mathfrak{g}^\theta$ has a rank one restricted root system Σ. Moreover, $\Sigma = \{\pm\tilde{\alpha}_i\}$ or $\Sigma = \{\pm\tilde{\alpha}_i, \pm 2\tilde{\alpha}_i\}$. There are four rank one cases that satisfy our assumptions. We list these cases below along with the desired element $\lambda \in P^+(\pi_\Theta)$ such that $\tilde{\lambda} = \omega_i'$.

$\mathfrak{g}$ is of type A_3 with $\pi = \{\alpha_1, \alpha_2, \alpha_3\}$, $\alpha_i = \alpha_2$
$\Theta(\alpha_2) = -\alpha_1 - \alpha_3 - \alpha_2$, and $\lambda = \omega_1$ or $\lambda = \omega_3$.

$\mathfrak{g}$ is of type B_r with $\pi = \{\alpha_1, \dots, \alpha_r\}$, $\alpha_i = \alpha_1$,
$\Theta(\alpha_1) = -\alpha_1 - 2\alpha_2 - \cdots - 2\alpha_r$, and $\lambda = \omega_r$.

$\mathfrak{g}$ is of type D_r with $\pi = \{\alpha_1, \dots, \alpha_r\}$, $\alpha_i = \alpha_1$
$\Theta(\alpha_1) = -\alpha_1 - 2\alpha_2 - \cdots - 2\alpha_{r-2} - \alpha_{r-1} - \alpha_r$,
and $\lambda = \omega_r$ or $\lambda = \omega_{r-1}$.

$\mathfrak{g}$ is of type C_r with $\pi = \{\alpha_1, \dots, \alpha_r\}$, $\alpha_i = \alpha_2$,
$\Theta(\alpha_i) = -\alpha_1 - \alpha_2 - 2\alpha_3 \cdots - 2\alpha_{r-1} - \alpha_r$, and $\lambda = \omega_1$. □

Set t equal to the rank of Σ. In Theorem 2.6, we find necessary and sufficient conditions for $\widetilde{P^+(\pi)}$ to equal $P^+(\Sigma)$. First, we take a look at a family of symmetric pairs for which this equality fails provided the rank of the restricted root system Σ is large enough. Consider first the special case when the symmetric pair $\mathfrak{g}, \mathfrak{g}^\theta$ is of type AII. It follows that $\pi_\Theta = \{\alpha_{2i+1} | \ 0 \leq i \leq t\}$ and the simple restricted roots are

$$\tilde{\alpha}_{2i} = (\alpha_{2i-1} + 2\alpha_{2i} + \alpha_{2i+1})/2 \tag{2.2}$$

for $1 \leq i \leq t$. Now assume that the Lie algebra $\mathfrak{g}$ is of classical type, $\mathfrak{g}$ contains a θ invariant Lie subalgebra $\mathfrak{r}$ such that $\mathfrak{r}, \mathfrak{r}^\theta$ is of type AII and the rank of the restricted root system associated to $\mathfrak{r}, \mathfrak{r}^\theta$ is $t-1$. It follows that the first $t-1$ simple restricted roots are given by the formula in (2.2). We list below the possibilities for $\mathfrak{g}, \mathfrak{g}^\theta$ under these assumptions and give the formula for the last simple restricted root. (It should be noted that in the cases below, if Σ is of type BC_t then $2\tilde{\alpha}_{2t}$ is also a restricted root.)

(2.3) $\mathfrak{g}, \mathfrak{g}^\theta$ is of type AII and Σ is of type A_t and $\tilde{\alpha}_{2t} = (\alpha_{2t-1} + 2\alpha_{2t} + \alpha_{2t+1})/2$.
(2.4) $\mathfrak{g}, \mathfrak{g}^\theta$ is of type CII(1), Σ is of type BC_t, $\tilde{\alpha}_{2t} = (\alpha_{2t-1} + 2\alpha_{2t} + 2\alpha_{2t+1} + \cdots + 2\alpha_{n-1} + \alpha_n)/2$.
(2.5) $\mathfrak{g}, \mathfrak{g}^\theta$ is of type CII(2), Σ is of type C_t, $\tilde{\alpha}_{2t} = \alpha_{2t} + \alpha_{2t-1}$.
(2.6) $\mathfrak{g}, \mathfrak{g}^\theta$ is of type DIII(1), Σ is of type C_t, $\tilde{\alpha}_{2t} = \alpha_{2t}$.
(2.7) $\mathfrak{g}, \mathfrak{g}^\theta$ is of type DIII(2), Σ is of type BC_t, $\tilde{\alpha}_{2t} = (\alpha_{2t+1} + \alpha_{2t-1} + \alpha_{2t})/2$.

Note that for each of the above cases, the set of positive restricted simple roots is just $\{\tilde{\alpha}_{2j} | \ 1 \leq j \leq t\}$. It follows that $P(\Sigma)$ is generated by the set of fundamental weights $\{\omega'_{2j} | \ 1 \leq j \leq t\}$. The next lemma describes the images of the fundamental weights inside $P(\pi)$ under $\tilde{}$ for the above five cases.

LEMMA 2.4. *Assume that $\mathfrak{g}, \mathfrak{g}^\theta$ satisfies the conditions of one of (2.3), (2.4), (2.5), (2.6) or (2.7). Set $t' = t$ if $\mathfrak{g}$ is not of type D_n and $t' = t-1$ if $\mathfrak{g}$ is of type D_n. We have*

(i) $\tilde{\omega}_1 = \omega'_2$
(ii) $\tilde{\omega}_{2i} = 2\omega'_{2i}$ *for all* $1 \leq i \leq t'$
(iii) $\tilde{\omega}_{2j+1} = \omega'_{2j} + \omega'_{2j+2}$ *for all* $1 \leq j \leq t'-1$
(iv) $\tilde{\omega}_{2t-1} = \omega'_{2t-2}$ *and* $\tilde{\omega}_{2t} = \omega'_{2t}$ *if* $\mathfrak{g}, \mathfrak{g}^\theta$ *is of type DIII(1)*
(v) $\tilde{\omega}_{2t} = \omega'_{2t} = \tilde{\omega}_{2t-1} = \tilde{\omega}_{2t+1}$ *if* $\mathfrak{g}, \mathfrak{g}^\theta$ *is of type DIII(2).*
(vi) $\tilde{\omega}_k$ *is a scalar multiple of* ω'_{2t} *for* $2t+1 \leq k \leq n$.

Moreover, if $\mathfrak{g}, \mathfrak{g}^\theta$ is of type CII(2) or DIII(1) then $n = 2t$ and otherwise $n > 2t$.

PROOF. Assertion (ii) follows from Lemma 2.1. Assertions (i) and (iii) follow from Lemma 2.3 and the description of the simple restricted roots given in (2.2), (2.4) and (2.5). Assertion (iv) follow from Lemma 2.3 and the formulas in (2.6) and (2.7). Assertion (v) follows from Lemma 2.1. Assertion (vi) as well as the last assertion of the lemma follow from the description of the simple restricted roots given in (2.2)-(2.7). □

The next lemma describes the image of $P^+(\pi)$ under $\tilde{}$ for five exceptional types of symmetric pairs. First, we specify the involution Θ and the restricted simple roots in each of these cases. Here, we number the roots $\alpha_1, \dots, \alpha_n$ as in [H, Section 13.2]. The first case is FII with underlying Lie algebra $\mathfrak{g}$ of type F_4. The involution Θ is the identity on the set $\pi_\Theta = \{\alpha_1, \alpha_2, \alpha_3\}$ and $\Theta(\alpha_4) = -\alpha_4 - 3\alpha_3 - 2\alpha_2 - \alpha_1$. In the second case $\mathfrak{g}, \mathfrak{g}^\theta$ is of type EIII with the Lie algebra $\mathfrak{g}$ of type E_6. The involution Θ is the identity on the set $\pi_\Theta = \{\alpha_3, \alpha_4, \alpha_5\}$ and satisfies $\Theta(\alpha_1) = -\alpha_3 - \alpha_4 - \alpha_5 - \alpha_6$, $\Theta(\alpha_2) = -\alpha_2 - 2\alpha_4 - \alpha_3 - \alpha_5$, and $\Theta(\alpha_6) = -\alpha_3 - \alpha_4 - \alpha_5 - \alpha_1$. The set of simple positive restricted roots for type EIII is $\{\tilde{\alpha}_1, \tilde{\alpha}_6\}$ where

$$\tilde{\alpha}_1 = (\alpha_1 + \alpha_3 + \alpha_4 + \alpha_5 + \alpha_6)/2$$
$$\tilde{\alpha}_2 = \alpha_2 + \alpha_4 + (\alpha_3 + \alpha_5)/2. \tag{2.8}$$

The three other types of symmetric pairs, EIV, EVII, and EIX, are very closely related to each other. In particular, the Lie algebra $\mathfrak{g}$ is of type E_6 for the symmetric pair EIV, of type E_7 for the symmetric pair EVII, and of type E_8 for the symmetric pair of type EIX. In each of these three cases, the involution Θ acts as the identity on $\pi_\Theta = \{\alpha_2, \alpha_3, \alpha_4, \alpha_5\}$, $\Theta(\alpha_i) = -\alpha_i$ for $7 \leq i \leq n$, $\Theta(\alpha_1) = -\alpha_1 - 2\alpha_3 - 2\alpha_4 - \alpha_2 - \alpha_5$ and $\Theta(\alpha_6) = -\alpha_6 - 2\alpha_5 - 2\alpha_4 - \alpha_2 - \alpha_3$. The set of simple restricted roots is $\{\tilde{\alpha}_1, \tilde{\alpha}_i | \ 6 \leq i \leq n\}$ where

$$\tilde{\alpha}_1 = \alpha_1 + \alpha_3 + \alpha_4 + (\alpha_2 + \alpha_5)/2$$
$$\tilde{\alpha}_6 = \alpha_6 + \alpha_5 + \alpha_4 + (\alpha_2 + \alpha_3)/2 \tag{2.9}$$
$$\tilde{\alpha}_i = \alpha_i \text{ for } i \geq 7.$$

LEMMA 2.5. (i) *If* $\mathfrak{g}, \mathfrak{g}^\theta$ *is of type FII, then*

$$\begin{aligned} \tilde{\omega}_4 &= 2\omega'_4 \\ \tilde{\omega}_1 &= 2\omega'_4 \\ \tilde{\omega}_2 &= 4\omega'_4 \\ \tilde{\omega}_3 &= 3\omega'_4 \end{aligned}$$

(ii) *If* $\mathfrak{g}, \mathfrak{g}^\theta$ *is of type EIII, then*

$$\begin{aligned} \tilde{\omega}_1 &= \tilde{\omega}_6 = \omega'_1 \\ \tilde{\omega}_3 &= \tilde{\omega}_5 = \omega'_1 + \omega'_2 \\ \tilde{\omega}_4 &= \omega'_1 + 2\omega'_2 \\ \tilde{\omega}_2 &= 2\omega'_2 \end{aligned}$$

(iii) *If* $\mathfrak{g}, \mathfrak{g}^\theta$ *is type EIV, EVII, or EIX then*

$$\begin{aligned} \tilde{\omega}_1 &= 2\omega'_1 & \tilde{\omega}_6 &= 2\omega'_6 \\ \tilde{\omega}_3 &= 2\omega'_1 + \omega'_6 & \tilde{\omega}_5 &= \omega'_1 + 2\omega'_6 \\ \tilde{\omega}_2 &= \omega'_1 + \omega''_6 & \tilde{\omega}_4 &= 2\omega'_1 + 2\omega'_6 \end{aligned}$$

Moreover, for $i \geq 7$, $\tilde{\omega}_i = \omega'_i$.

PROOF. The first line of equalities in (i), (ii), and (iii) follow from Lemma 2.1. The remaining equalities are straightforward calculations. □

Using Lemma 2.4 and Lemma 2.5, we are now ready to provide necessary and sufficient condition for $\widetilde{P^+(\pi)}$ to be equal to $P^+(\Sigma)$.

THEOREM 2.6. $\widetilde{P^+(\pi)} = P^+(\Sigma)$ *if and only if* $\mathfrak{g}, \mathfrak{g}^\theta$ *is not of type FII, EIII, EIV, EVII, EIX, or CII(2), and* $\mathfrak{g}$ *does not contain a* θ *invariant Lie subalgebra* $\mathfrak{r}$ *of rank greater than or equal to* 7 *such that* $\mathfrak{r}, \mathfrak{r}^\theta$ *is of type AII.*

PROOF. First assume that $\mathfrak{g}, \mathfrak{g}^\theta$ is not of type FII, EIII, EIV, EVII, EIX, or CII(2), and $\mathfrak{g}$ does not contain a θ invariant Lie subalgebra $\mathfrak{r}$ of rank greater than or equal to 7 such that $\mathfrak{r}, \mathfrak{r}^\theta$ is of type AII. If $\mathfrak{g}, \mathfrak{g}^\theta$ is not of type EVI, it is straightforward to check that each simple root of $\mathfrak{g}$ satisfies the conditions of Lemma 2.3. On the other hand, if $\mathfrak{g}, \mathfrak{g}^\theta$ is of type EVI, then $\tilde{\omega}_1 = \omega'_1$, $\tilde{\omega}_3 = \omega'_3$, $\tilde{\omega}_2 = \omega'_4$ and $\tilde{\omega}_7 = \omega'_6$. (Here, we are numbered the simple roots as in [H, Section 11.4].) Hence $\widetilde{P^+(\pi)} = P^+(\Sigma)$ in this case as well.

Suppose that $\mathfrak{g}, \mathfrak{g}^\theta$ is of type FII, EIII, EIV, EVII, or EIX. Using Lemma 2.5, it is straightforward to check that the set $\{\tilde{\omega}_i | \ 1 \leq i \leq n\}$ generates a proper subset of $P^+(\Sigma)$.

Now assume that $\mathfrak{g}, \mathfrak{g}^\theta$ is one of the types listed in (2.3) - (2.7). Assume further that either $\mathfrak{g}, \mathfrak{g}^\theta$ is of type CII(2) or $\mathfrak{g}$ contains a θ invariant Lie subalgebra $\mathfrak{r}$ of rank greater than or equal to 7 such that $\mathfrak{r}, \mathfrak{r}^\theta$ is of type AII. These assumptions and Lemma 2.4 yield that

$$\begin{aligned} \tilde{\omega}_1 &= \omega'_2 & \tilde{\omega}_4 &= 2\omega'_4 \\ \tilde{\omega}_2 &= 2\omega'_2 & \tilde{\omega}_5 &= \omega'_4 + \omega'_6 \\ \tilde{\omega}_3 &= \omega'_2 + \omega'_4 & \tilde{\omega}_6 &= 2\omega'_6 \end{aligned}$$

Note that ω'_4 cannot be written as a linear combination of the $\tilde{w}_i$, for $1 \leq i \leq 6$, with nonnegative coefficients. Lemma 2.4 ensures that for all $k \geq 7$, $\tilde{\omega}_k$ is a linear combination of elements in the set $\{\omega'_j | \ j \notin \{2,4\}\}$. In particular, ω'_4 is not in $\widetilde{P^+(\pi)}$. □

Consider the case when $P^+(\Sigma) = \widetilde{P^+(\pi)}$. It follows that $P(\Sigma) = \widetilde{P(\pi)}$. We show that this second equality holds for all symmetric pairs.

THEOREM 2.7. *The set* $\widetilde{P(\pi)}$ *equals* $P(\Sigma)$ *for all irreducible symmetric pairs.*

PROOF. By Lemma 2.2 we only need to show that $P(\Sigma)$ is a subset of $\widetilde{P(\pi)}$. In particular, it is sufficient to show that the fundamental weights associated to Σ are all contained in $\widetilde{P(\pi)}$. The comments preceding the theorem and Theorem 2.6 allow us to reduce to the following cases:

(i) $\mathfrak{g}, \mathfrak{g}^\theta$ is one of the exceptional symmetric pairs FII, EIII, EIV, EVII, or EIX
(ii) $\mathfrak{g}$ is classical and $\mathfrak{g}$ contains a θ invariant Lie subalgebra $\mathfrak{r}$ such that $\mathfrak{r}, \mathfrak{r}^\theta$ is of type AII and the rank of Σ is greater than or equal to 4.

Assume $\mathfrak{g}, \mathfrak{g}^\theta$ satisfies the conditions of Case (ii). Note that the fundamental weights associated to the root system Σ are just the elements in the set $\{\omega'_{2j} | \ 1 \leq j \leq t\}$ where t is the rank of Σ. By Lemma 2.4(i) $\widetilde{P(\pi)}$ contains ω'_2. Moreover, by Lemma 2.4 (iii) $\omega'_{2i} + \omega'_{2(i+1)}$ is contained in $\widetilde{P(\pi)}$ for $1 \leq i \leq t-2$. Using an inductive argument, we may assume that $\omega'_{2j} \in \widetilde{P(\pi)}$ where j is an integer between 1 and $t-3$. Then $\omega'_{2(j+1)} = (\omega'_{2(j+1)} + \omega'_{2j}) - \omega'_{2j}$ and so $\omega'_{2(j+1)} \in \widetilde{P(\pi)}$. Thus $\omega'_2, \ldots, \omega'_{2(t-1)}$ are all contained in $\widetilde{P(\pi)}$. Now if $\mathfrak{g}$ is not of type D_n, then by Lemma 2.4 (iii), $\omega'_{2(t-1)} + \omega'_{2t}$ is also in $\widetilde{P(\pi)}$. Thus since $\omega'_{2(t-1)} \in \widetilde{P(\pi)}$, so is ω'_{2t}. On the other hand if $\mathfrak{g}$ is of type D_n, then by Lemma 2.4(iv) and (v), $\omega'_{2t} \in \widetilde{P(\pi)}$. This completes the proof in Case (ii).

Suppose $\mathfrak{g}, \mathfrak{g}^\theta$ is of type FII. By Lemma 2.5, we see that $\omega' = \tilde{\omega}_2 - \tilde{\omega}_3$. Thus $\widetilde{P(\pi)}$ contains all integer multiples of ω' and the theorem follows in this case. Consider the cases when $\mathfrak{g}, \mathfrak{g}^\theta$ is of type EIV, EVII, or EIX. A straightforward computation using Lemma 2.5 shows that $\omega'_1 = \tilde{\omega}_5 - \tilde{\omega}_6$ and $\omega'_6 = \tilde{\omega}_3 - \tilde{\omega}_1$. Furthermore, by Lemma 2.5, we have $\tilde{\omega}_i = \omega'_i$ for $i \geq 7$. Thus $\widetilde{P(\pi)}$ contains all the fundamental restricted roots and the theorem follows for these cases. Now assume $\mathfrak{g}, \mathfrak{g}^\theta$ is of type EIII. By Lemma 2.5, we have $\tilde{\omega}_1 = \omega'_1$. One further checks using Lemma 2.5 that $\omega'_2 = \tilde{\omega}_3 - \tilde{\omega}_1$ which completes the proof of the theorem. □

CHAPTER 3

Twisted Weyl Group Actions

This section is a study of the so-called dotted actions of both W and W_Θ. The dotted actions of W appear naturally in the analysis of the image of the center of $\check{U}$ under the ordinary Harish-Chandra map. After reviewing the use of the dotted actions associated to the center, we draw connections between the dotted actions of W and the dotted actions of W_Θ.

The Weyl group associated to $\mathfrak{g}$ acts on the vector space $\mathfrak{h}^*$. The dotted action of W on $\mathfrak{h}^*$ is a slight twist of this action. In particular, recall that ρ is the half sum of the positive roots for $\mathfrak{g}$. Set

$$w \cdot \lambda = w(\lambda + \rho) - \rho$$

for all $\lambda \in \mathfrak{h}^*$ and $w \in W$. The ordinary action of W on $\mathfrak{h}^*$ induces an action of W on $\check{T}$ and hence on $\check{U}^0$. Once again, this action can be twisted to obtain a dotted action of W on $\check{U}^0$ such that

$$w \cdot q^{(\rho,\mu)}\tau(\mu) = q^{(\rho,w\mu)}\tau(w\mu) \tag{3.1}$$

for all $w \in W$ and $\mu \in \mathfrak{h}^*$. These two dotted actions are connected by the following straightforward to check fact

$$z^{w\cdot\lambda}(a) = z^{\lambda}(w^{-1} \cdot a) \tag{3.2}$$

for all $\lambda \in \mathfrak{h}^*$ and $a \in \check{U}^0$.

We review here the definition of the ordinary quantum Harish-Chandra projection. Using the triangular decomposition of $\check{U}$ (see (1.10)), $\check{U}$ admits the following direct sum decomposition.

$$\check{U} = \check{U}^0 \oplus (G_+^-\check{U} + \check{U}U_+^+). \tag{3.3}$$

Let $\mathcal{P}$ denote the projection of $\check{U}$ onto $\check{U}^0$ using this decomposition. This is a quantum version of the classical Harish-Chandra map associated to enveloping algebras of semisimple Lie algebras. Just as in the classical case, the ordinary quantum Harish-Chandra map induces an isomorphism between the center of $\check{U}$ and twisted Weyl group invariants of the Cartan part. More precisely, we have the following description of the center under the ordinary Harish-Chandra map ([Jo, Lemma 7.1.17 and 7.1.25]).

THEOREM 3.1. *The ordinary Harish-Chandra map $\mathcal{P}$ defines an isomorphism from $Z(\check{U})$ onto $\mathcal{C}[\tau(2\lambda)|\ \lambda \in P(\pi)]^{W\cdot}$.*

Let $\mathfrak{a}^*$ be the vector subspace of $\mathfrak{h}^*$ equal to the eigenspace of Θ with eigenvalue -1. Note that $\mathfrak{a}^*$ can be identified with the set containing all $\tilde{\alpha}$ with $\alpha \in \mathfrak{h}^*$ (see (1.4)). We define a group $\check{\mathcal{A}}$ that can be viewed as a quantum version of $\mathfrak{a}^*$, or, more precisely, an analog of the lattice of integral restricted weights contained in $\mathfrak{a}^*$. In particular, let $\check{A}$ be the group defined by

$$\check{A} = \{\tau(\mu) | \ \mu \in P(\Sigma)\}.$$

By Theorem 2.7, we have that $\{\tau(\tilde{\beta}) | \beta \in P(\pi)\} = \check{A}$. Using the definition of $\tilde{\beta}$ in (1.4), we see that $2\tilde{\beta} \in Q(\pi)$ whenever $\beta \in Q(\pi)$. In particular, we cannot expect $\check{A}$ to be a subgroup of $\check{T}$. However, $\check{A}$ is a subgroup of the larger group $\{\tau(\mu) | \ 2\mu \in P(\pi)\}$ containing $\check{T}$.

The dotted action of the restricted Weyl group W_Θ on $\mathfrak{a}^*$ and $\mathcal{C}[\check{A}]$ is defined using very similar formulas to the ones above. To avoid confusion between the two dotted actions, we use a different "dot" to denote this second action. In particular, note that $\tilde{\rho} = (\rho + \Theta(-\rho))/2$. Given $w \in W_\Theta$, $\lambda \in \mathfrak{a}^*$, and $\tau(\mu) \in \check{A}$, set

$$w \circ \lambda = w(\lambda + \tilde{\rho}) - \tilde{\rho}$$

and

$$w \circ q^{(\tilde{\rho},\mu)}\tau(\mu) = q^{(\tilde{\rho},w\mu)}\tau(w\mu).$$

Just as for the dotted action of the ordinary Weyl group W, a routine computation shows that

$$z^{w\circ\lambda}(a) = z^{\lambda}(w^{-1} \circ a)$$

for all $\lambda \in \mathfrak{a}^*$ and a in $\mathcal{C}[\check{A}]$.

As explained in the discussion preceding Lemma 2.2, an element $\tilde{w}$ in W_Θ can be lifted to w in W such that $\tilde{w}$ and w act the same on $Q(\Sigma)$. It follows that w restricts to an automorphism of the root system Σ. Moreover, since both Θ and W preserve the Cartan inner product this condition ensures that w restricts to an automorphism of the root system generated by π_Θ. The next lemma will be used to relate the two dotted actions of W with the corresponding dotted actions of W_Θ.

LEMMA 3.2. *Given $\tilde{w} \in W_\Theta$, we can find $w \in W$ such that the restriction of w to Σ is equal to $\tilde{w}$ and $w(\rho - \tilde{\rho}) = \rho - \tilde{\rho}$.*

PROOF. It is sufficient to prove the lemma for those elements $\tilde{w} \in W_\Theta$ that are reflections corresponding to simple restricted roots. In particular, fix i with $\alpha_i \in \pi^*$. Let $\tilde{s}_i$ be the reflection in W_Θ corresponding to $\tilde{\alpha}_i$. Let s_i be an element in W such that s_i and $\tilde{s}_i$ agree on Σ. By the discussion preceding the lemma, s_i restricts to an automorphism of the root system associated to π_Θ.

Let W' denote the subgroup of W generated by the reflections associated to roots in π_Θ. In particular, W' can be viewed as the Weyl group associated to the root system generated by π_Θ. Since s_i acts as an automorphism on this root system, we can find $u \in W'$ such that $s_i u$ is a diagram automorphism on π_Θ. Now $(\tilde{\beta}, \alpha_j) = 0$ for all $\alpha_j \in \pi_\Theta$ and $\tilde{\beta} \in \Sigma$. Thus reflections associated to simple roots in π_Θ act trivially on Σ. It follows that elements of W', and in particular u, acts as the identity on Σ. Hence $s_i u$ also restricts to the reflection $\tilde{s}_i$ on Σ.

Let ρ_Θ denote the half sum of the positive roots in the root system generated by π_Θ. Since π_Θ is a root system with the same inner product as π, we must have ([H, Corollary 10.2]) $(\rho_\Theta, \alpha_i) = (\alpha_i, \alpha_i)/2 = (\rho, \alpha_i)$ for all $\alpha_i \in \pi_\Theta$. The fact that $s_i u$ is a diagram automorphism on π_Θ ensures that $s_i u \rho_\Theta = \rho_\Theta$. Now $(\tilde{\rho}, \alpha_j) = 0$ for all $\alpha_j \in \pi_\Theta$. Hence $(\rho - \tilde{\rho}, \alpha_j) = (\rho, \alpha_j)$ for all $\alpha_j \in \pi_\Theta$. Therefore

$$(\rho - \tilde{\rho}, \beta) = (\rho_\Theta, \beta) \text{ for all } \beta \in Q(\pi_\Theta). \tag{3.4}$$

Note that

$$\pi_\Theta \cup \{\alpha_i - \alpha_{\mathrm{p}(i)} | \alpha_i \in \pi^* \text{ and } i \neq \mathrm{p}(i)\}$$

is a linearly independent subset of $\mathfrak{h}^*$. Since $\rho - \tilde{\rho}$ is Θ invariant, we can write it as a linear combination of elements in the above set. Recall that p is a permutation on $\{1, \ldots, n\}$ that induces a diagram automorphism of π. Temporarily call this diagram automorphism d. Note that d extends to a linear isomorphism of $\mathfrak{h}^*$ which we also refer to as d. Since ρ is the half sum of the positive roots, we have $d\rho = \rho$. Furthermore, (1.19) and (1.20) ensure that $d\Theta(\alpha_j) = \Theta(\alpha_{\mathrm{p}(j)})$ for all $\alpha_j \in \pi^*$. Hence $d\tilde{\rho} = \tilde{\rho}$. Since $d(\alpha_i - \alpha_{\mathrm{p}(i)}) = -\alpha_i + \alpha_{\mathrm{p}(i)}$, the coefficient of $\alpha_i - \alpha_{\mathrm{p}(i)}$ in $\rho - \tilde{\rho}$ must be zero. Therefore $\rho - \tilde{\rho}$ is in $Q(\pi_\Theta)$. By (3.4), we have that $\rho - \tilde{\rho} = \rho_\Theta$. Hence $s_i u(\rho - \tilde{\rho}) = \rho - \tilde{\rho}$. The lemma follows by setting $w = s_i u$. □

Let $\check{T}_\Theta$ be the group defined by

$$\check{T}_\Theta = \{\tau((\mu + \Theta(\mu))/2) |\ \mu \in P(\pi)\}.$$

Note that $\check{T}_\Theta$ is a subgroup of the extension $\{\tau(\mu) |\ 2\mu \in P(\pi)\}$ of $\check{T}$. Note further that $\tau(\mu) = \tau(\tilde{\mu})\tau((\mu + \Theta(\mu))/2) \in \check{\mathcal{A}}\check{T}_\Theta$ for all $\mu \in P(\pi)$. Hence $\check{T}$ is a subgroup of $\check{\mathcal{A}}\check{T}_\Theta$. This implies the following inclusion:

$$\check{U}^0 \subseteq \mathcal{C}[\check{\mathcal{A}}] \oplus \mathcal{C}[\check{\mathcal{A}}]\mathcal{C}[\check{T}_\Theta]_+. \tag{3.5}$$

Let $\tilde{\mathcal{P}}$ be the projection of $\check{U}^0$ onto $\mathcal{C}[\check{\mathcal{A}}]$ using this direct sum decomposition. Note that $\tilde{\mathcal{P}}$ sends an element $\tau(\mu)$ in $\check{T}$ to $\tau(\tilde{\mu})$.

Let $\mathcal{A}$ be the subgroup of $\check{\mathcal{A}}$ defined by

$$\mathcal{A} = \{\tau(2\mu) |\ \mu \in P(\Sigma)\}.$$

(It should be noted that this definition of $\mathcal{A}$ differs from the definition of $\mathcal{A}$ in [L5].) Using Theorem 2.7, we see that $\mathcal{A} = \{\tau(2\tilde{\mu}) |\ \mu \in P(\pi)\}$. Recall that Θ is an automorphism of the root system Δ generated by π. Thus the definition of $\tilde{\mu}$ given in (1.4) ensures that $\mathcal{A}$ is a subgroup of $\check{T}$. The next lemma shows that $\tilde{\mathcal{P}} \circ \mathcal{P}(Z(\check{U}))$ is invariant under the dotted action of the restricted Weyl group W_Θ. Eventually, this result will be used to establish the dotted W_Θ invariance of the image of $Z(\check{U})$ and $\check{U}^B$ under the quantum Harish-Chandra map associated to a quantum symmetric pair.

THEOREM 3.3. *Suppose that $f \in \check{U}^0$ and f is invariant under the dotted action of W. Then $\tilde{\mathcal{P}} \circ \mathcal{P}(f)$ is a dotted W_Θ invariant element of $\mathcal{C}[\check{\mathcal{A}}]$. Moreover, $\tilde{\mathcal{P}} \circ \mathcal{P}(Z(\check{U}))$ is a subring of $\mathcal{C}[\mathcal{A}]^{W_\Theta \circ}$.*

PROOF. Suppose that $\tilde{w} \in W_\Theta$. By Lemma 3.2, we can choose w in W such that $\tilde{w}^{-1}$ and w^{-1} agree on Σ and $w^{-1}(\rho - \tilde{\rho}) = \rho - \tilde{\rho}$. Now suppose that $\lambda \in P(\pi)$. We have $(\beta, \tilde{\lambda}) = (\beta, \lambda)$ for all $\beta \in \mathfrak{a}^*$. In particular, $(\beta, \tilde{\lambda} - \lambda) = 0$ for all $\beta \in \mathfrak{a}^*$. Note that $w^{-1}\tilde{\rho} = \tilde{w}^{-1}\tilde{\rho}$ is an element of $\mathfrak{a}^*$. It follows that

$$\begin{aligned}(\rho, \tilde{\lambda} - \lambda + w\lambda - w\tilde{\lambda}) &= (\rho, \tilde{\lambda} - \lambda) + (w^{-1}\rho, \lambda - \tilde{\lambda}) \\ &= (\rho - w^{-1}\rho, \tilde{\lambda} - \lambda) \\ &= (\tilde{\rho} - w^{-1}\tilde{\rho}, \tilde{\lambda} - \lambda) = 0.\end{aligned}$$

Hence

$$
\begin{aligned}
w \cdot \tau(\lambda) &= q^{(\rho,-\lambda)} q^{(\rho,w\lambda)} \tau(w\lambda) \\
&= q^{(\rho,-\tilde{\lambda})} q^{(\rho,w\tilde{\lambda})} q^{(\rho,-\lambda+\tilde{\lambda}+w\lambda-w\tilde{\lambda})} \tau(w\lambda) \\
&= q^{(\rho,-\tilde{\lambda})} q^{(\rho,w\tilde{\lambda})} \tau(w\tilde{\lambda}) \tau(w(\lambda+\Theta(\lambda))) \\
&= q^{(\tilde{\rho},-\tilde{\lambda})} q^{(\tilde{\rho},w\tilde{\lambda})} \tau(w\tilde{\lambda}) + q^{(\rho,-\tilde{\lambda})} q^{(\rho,w\tilde{\lambda})} \tau(w\tilde{\lambda})(\tau(w((\lambda+\Theta(\lambda))/2)) - 1).
\end{aligned}
$$

Thus $\tilde{\mathcal{P}}(w \cdot \tau(\lambda)) = \tilde{w} \circ \tau(\tilde{\lambda})$ for all $\tau(\lambda) \in \check{T}$. Since $w \cdot f = f$, it follows that $\tilde{w} \circ \tilde{\mathcal{P}}(f) = \tilde{\mathcal{P}}(f)$.

By Theorem 3.1, $\mathcal{P}(Z(\check{U}))$ is a subring of $\mathcal{C}[\tau(2\mu)|\ \mu \in P(\pi)]$. The second assertion now follows from the fact that $\mathcal{C}[\tau(2\tilde{\mu})|\ \mu \in P(\pi)]$ is contained in the group algebra of $\mathcal{A}$. $\square$

CHAPTER 4

The Harish-Chandra Map

In this section, we construct a quantum Harish-Chandra map associated to the symmetric pair $\mathfrak{g}, \mathfrak{g}^\theta$. The procedure involves four tensor product decompositions of the quantized enveloping algebra with respect to a subalgebra $B \in \mathcal{B}$. The results establishing these decompositions parallel the material presented in [L5, Section 2]. However, the theorems in [L5, Section 2] are only proved for the standard analogs of $U(\mathfrak{g}^\theta)$ which have a reduced restricted root system Σ. In particular, we first generalize [L5, Lemma 2.1] and [L5, Theorem 2.2] so that it applies to all coideal subalgebras in $\mathcal{B}$ for any irreducible symmetric pair $\mathfrak{g}, \mathfrak{g}^\theta$. The quantum Harish-Chandra map associated to an algebra $B \in \mathcal{B}$ is then defined using these decompositions. The end of the section focuses on the restriction of this map to the $(\mathrm{ad}_r\ B)$ invariants of $\check{U}$. The tensor product decompositions of this section are also used in Section 5 to generalize the notion of radial components defined in [L5, Section 3].

We introduce some notation from [L5, Section 2]. Set $\mathcal{M}^+ = \mathcal{M} \cap U^+$ and $\mathcal{M}^- = \mathcal{M} \cap G^-$. Recall (Section 1) that ad is used to denote the left adjoint action of U on itself. Let N^+ be the subalgebra of U^+ generated by the $(\mathrm{ad}\ \mathcal{M}^+)$ module $(\mathrm{ad}\ \mathcal{M}^+)\mathcal{C}[x_i|\alpha_i \notin \pi_\Theta]$. Similarly, let N^- be the subalgebra of G^- generated by the $(\mathrm{ad}\ \mathcal{M}^-)$ module $(\mathrm{ad}\ \mathcal{M}^-)\mathcal{C}[y_it_i|\alpha_i \notin \pi_\Theta]$.

Given $\beta \in Q(\pi)$ and a subspace S of U, we write $S_{\beta,r}$ for the sum of weight subspaces $S_{\beta'}$ of S with $\tilde{\beta}' = \tilde{\beta}$. For each $\gamma \in Q^+(\pi)$, set $T(\gamma)$ equal to the subset of T consisting of those $\tau(\eta)$ where

$$\eta \in \sum_{\alpha_i \in \pi \setminus \pi_\Theta} \mathbf{N}2\alpha_i + \sum_{\alpha_j \in \mathcal{S}} \mathbf{N}\alpha_j \quad \text{and} \quad 0 \leq \eta \leq \gamma. \tag{4.1}$$

The following extends [L5, Lemma 2.1] to the nonstandard analogs and to those symmetric pairs with nonreduced restricted root system. The lemma below is slightly stronger than the original version given in [L5]. As a result, we give a complete proof here.

LEMMA 4.1. *For each $B \in \mathcal{B}$, all $\beta, \gamma \in Q^+(\pi)$, and $Y \in U^+_\gamma G^-_{-\beta}$, we have*

$$(i)\quad Y \in N^+_{\beta+\gamma,r}B + \sum_{\{\beta' |\ \tilde{\beta}' < \tilde{\beta}+\tilde{\gamma}\}} N^+_{\beta'}T(\beta+\gamma-\beta')B$$

$$(ii)\quad Y \in BN^+_{\beta+\gamma,r} + \sum_{\{\beta' |\ \tilde{\beta}' < \tilde{\beta}+\tilde{\gamma}\}} BT(\beta+\gamma-\beta')N^+_{\beta'}$$

$$(iii)\quad Y \in N^-_{-\beta-\gamma,r}B + \sum_{\{\beta' |\ \tilde{\beta}' < \tilde{\beta}+\tilde{\gamma}\}} N^-_{-\beta'}T(\beta+\gamma-\beta')B$$

$$(iv)\quad Y \in BN^-_{-\beta-\gamma,r} + \sum_{\{\beta' \mid \tilde{\beta}' < \tilde{\beta}+\tilde{\gamma}\}} BT(\beta+\gamma-\beta')N^-_{-\beta'}$$

PROOF. We consider here the proof of the first assertion; the remaining assertions follows using a similar argument. We will be using the fact that $\tilde{\lambda} = 0$ for all $\lambda \in Q^+(\pi_\Theta)$ to switch back and forth between conditions on restricted weights and weights in $Q(\pi)$. Let $B \in \mathcal{B}$. Now any Hopf algebra automorphism which fixes elements of T also restricts to an automorphism of both N^- and N^+. Hence we may assume that B is one of the coideal algebras $B_\theta, B_{\theta,0,d}$, or $B_{\theta,s,1}$. In particular, we may assume that B is generated by $\mathcal{M}, T_\Theta$ and the B_i for $\alpha_i \notin \pi_\Theta$ where $B_i = y_it_i + d_i\tilde{\theta}(y_i)t_i + s_it_i$ (see Section 1).

Fix $\beta \in Q^+(\pi)$. Note that N^+ is an ad $\mathcal{M}^+$ module. Suppose that $\alpha_i \in \pi_\Theta$. Using the formulas for the left adjoint action given in (1.15), we can write $x_ia = (\text{ad } x_i)a + t_iat_i^{-1}x_i$ for all $a \in U$. It follows that $x_iN^+_\beta \subseteq N^+_{\beta+\alpha_i} + N^+_\beta x_i$. Hence

$$\mathcal{M}^+N^+_\beta \subseteq \sum_{\mu\in\beta+Q^+(\pi_\Theta)} N^+_\mu\mathcal{M}^+. \tag{4.2}$$

On the other hand, if $\alpha_i \notin \pi_\Theta$, we have that $x_i \in N^+$ and hence $x_iN^+_\beta \subseteq N^+_{\beta+\alpha}$. Therefore

$$U^+_\gamma N^+_\beta \subseteq \sum_{\mu\,\in\,\beta+\gamma+Q^+(\pi_\Theta)} N^+_\mu\mathcal{M}^+$$

for all γ and β in $Q^+(\pi)$. Thus it is sufficient to prove (i) when Y is an element of $G^-_{-\beta}$ and $\gamma = 0$.

Recall (1.6) that $x_jy_i - y_ix_j = \delta_{ij}(t_i - t_i^{-1})(q_i - q_i^{-1})^{-1}$. Hence

$$G^-_{-\beta}x_i \subseteq x_iG^-_{-\beta} + G^-_{-\beta+\alpha_i} + G^-_{-\beta+\alpha_i}t_i^2$$

for all i such that $1 \le i \le n$. It follows that

$$G^-_{-\beta}x_i \subseteq x_iG^-_{-\beta} + G^-_{-\beta+\alpha_i}T_\Theta \text{ for all } \alpha_i \in \pi_\Theta \tag{4.3}$$

and

$$G^-_{-\beta}x_i \subseteq x_iG^-_{-\beta} + G^-_{-\beta+\alpha_i} + G^-_{-\beta+\alpha_i}t_i^2 \text{ for all } \alpha_i \notin \pi_\Theta. \tag{4.4}$$

Let Y be an element of $G^-_{-\beta}$. Without loss of generality, we may assume that Y is a monomial of the form $y_{i_1}t_{i_1}\cdots y_{i_m}t_{i_m}$. If $m = 0$, then Y is just a scalar and the conclusions of the lemma easily hold. Assume $m > 0$. We proceed by induction on m. In particular, we assume that all monomials in the y_it_i of length less than or equal to $m-1$ satisfy (i).

Suppose that $\alpha_{i_m} \in \pi_\Theta$. It follows that $y_{i_m}t_{i_m} \in B$. Applying the inductive hypothesis to $y_{i_1}t_{i_1}\cdots y_{i_{m-1}}t_{i_{m-1}}$ yields that (i) holds for Y. Thus we may assume that $\alpha_{i_m} \notin \pi_\Theta$.

Using the description of the elements B_i above, we can write

$$\begin{aligned} Y =& y_{i_1}t_{i_1}\cdots y_{i_{m-1}}t_{i_{m-1}}(B_{i_m}) \\ &- d_{i_m}y_{i_1}t_{i_1}\cdots y_{i_{m-1}}t_{i_{m-1}}\tilde{\theta}(y_{i_m})t_{i_m} \\ &- s_{i_m}y_{i_1}t_{i_1}\cdots y_{i_{m-1}}t_{i_{m-1}}t_{i_m}. \end{aligned}$$

Note that

$$N^+_{\beta-\alpha_{i_m},r}B + \sum_{\{\beta' \mid \tilde{\beta}' < \tilde{\beta}-\tilde{\alpha}_{i_m}\}} N^+_{\beta'}T(\beta-\alpha_{i_m}-\beta')B$$

is a subset of

$$\sum_{\{\beta' \mid \tilde{\beta}' < \tilde{\beta}\}} N^+_{\beta'} T(\beta - \beta')B.$$

Hence, by the inductive hypothesis, we see that

$$y_{i_1} t_{i_1} \cdots y_{i_{m-1}} t_{i_{m-1}} B_{i_m}$$

is contained in the right hand side of (i).

Note that $s_{i_m} y_{i_1} t_{i_1} \cdots y_{i_{m-1}} t_{i_{m-1}} t_{i_m}$ is just a nonzero scalar multiple of the term $s_{i_m} t_{i_m} y_{i_1} t_{i_1} \cdots y_{i_{m-1}} t_{i_{m-1}}$. Another application of the inductive hypothesis yields that

$$\begin{aligned} t_{i_m} y_{i_1} t_{i_1} \cdots y_{i_{m-1}} t_{i_{m-1}} &\in t_{i_m} \sum_{\{\beta' \mid \tilde{\beta}' < \tilde{\beta}\}} N^+_{\beta'} T(\beta - \alpha_{i_m} - \beta')B \\ &\subseteq \sum_{\{\beta' \mid \tilde{\beta}' < \tilde{\beta}\}} N^+_{\beta'} T(\beta - \beta')B. \end{aligned}$$

In particular, $s_{i_m} y_{i_1} t_{i_1} \cdots y_{i_{m-1}} t_{i_{m-1}} t_{i_m}$ is also an element of the right hand side of (i).

We now adapt the argument of [L5, Lemma 2.2] to show that the remaining term $d_{i_m} y_{i_1} t_{i_m} \cdots y_{i_{m-1}} t_{i_{m-1}} \tilde{\theta}(y_{i_m}) t_{i_m}$ of Y is in the right hand side of (i). We see from the description of $\tilde{\theta}$ in (1.19) and (1.20) that $\tilde{\theta}(y_{i_m}) t_{i_m} \in \mathcal{M}^+ x_{\mathrm{p}(i_m)} \mathcal{M}^+ T_\Theta$. Hence

$$d_{i_m} y_{i_1} t_{i_m} \cdots y_{i_{m-1}} t_{i_{m-1}} \tilde{\theta}(y_{i_m}) t_{i_m} \in G^-_{-\beta+\alpha_{i_m}} \mathcal{M}^+ x_{\mathrm{p}(i_m)} \mathcal{M}^+ T_\Theta.$$

By (4.2), (4.3) and (4.4), we have

$$\begin{aligned} G^-_{-\beta+\alpha_{i_m}} \mathcal{M}^+ x_{\mathrm{p}(i_m)} \mathcal{M}^+ T_\Theta &\subseteq \sum_{\eta} \sum_{\mu} N^+_\eta G^-_{-\mu} \mathcal{M}^+ T_\Theta \\ &\quad + \sum_{\mu'} G^-_{-\mu'} \mathcal{M}^+ (\mathcal{C}[T_\Theta] + t^2_{\mathrm{p}(i)} \mathcal{C}[T_\Theta]). \end{aligned}$$

Here, the first term is a sum taken over values of η in the set $\alpha_{i_m} + Q^+(\pi_\Theta)$ and μ in $Q^+(\pi)$ such that $-\mu \in -\beta + \alpha_{i_m} + Q^+(\pi_\Theta)$. The second term is a sum taken over values of μ' in $Q^+(\pi)$ such that $-\mu' \in -\beta + \alpha_{i_m} + \alpha_{\mathrm{p}(i_m)} + Q^+(\pi_\Theta)$. Recall that α_{i_m} is not in π_Θ. It follows that the choices of μ and μ' guarantee that elements of $G^-_{-\mu}$ and $G^-_{-\mu'}$ can be written as linear combinations of monomials in the $y_i t_i$ with length strictly less than m. The lemma now follows from the reduction to elements in G^- described at the beginning of the proof and induction on m. □

Let T' be the subgroup of T generated by t_i for $\alpha_i \in \pi^*$. The next result provides four tensor product decompositions of U. This generalizes [L5, Theorem 2.2] which is proved for the subset of standard analogs in $\mathcal{B}$. Though the proof here is quite similar to the argument in [L5], one of the antiautomorphisms used in [L5] does not work in this more general setting.

THEOREM 4.2. *For all $B \in \mathcal{B}$, there are vector space isomorphisms via the multiplication map*

(i) $N^+ \otimes \mathcal{C}[T'] \otimes B \cong U$
(ii) $B \otimes \mathcal{C}[T'] \otimes N^+ \cong U$
(iii) $N^- \otimes \mathcal{C}[T'] \otimes B \cong U$
(iv) $B \otimes \mathcal{C}[T'] \otimes N^- \cong U$

PROOF. Fix $B \in \mathcal{B}$. By [L3, Section 6], the multiplication map induces an isomorphism

$$U^+ \cong \mathcal{M}^+ \otimes N^+. \tag{4.5}$$

Switching the order of $\mathcal{M}^+$ and N^+, the same argument shows that multiplication induces an isomorphism

$$U^+ \cong N^+ \otimes \mathcal{M}^+. \tag{4.6}$$

Note that $T = T' \times T_\Theta$. It follows that

$$U^0 \cong \mathcal{C}[T_\Theta] \otimes \mathcal{C}[T'] \tag{4.7}$$

where the isomorphism comes from the multiplication map. The triangular decomposition of U (1.10), (4.5), and (4.7) imply that multiplication induces an isomorphism

$$U \cong G^- \otimes \mathcal{M}^+ \otimes \mathcal{C}[T_\Theta] \otimes \mathcal{C}[T'] \otimes N^+.$$

A similar argument using (4.6) yields that multiplication also induces an isomorphism

$$U \cong N^+ \otimes \mathcal{C}[T'] \otimes \mathcal{C}[T_\Theta] \otimes \mathcal{M}^+ \otimes G^-.$$

As explained in [L5] (see [L5,(2.9)] and subsequent discussion), the lowest weight term of an element $b \in B$ written as a sum of weight vectors is in $G^-\mathcal{M}^+T_\Theta$. By [L3, Section 6 and (6.2)] and [Ke], we have $G^-\mathcal{M}^+T_\Theta = N^-\mathcal{M}T_\Theta = \mathcal{M}T_\Theta N^- = \mathcal{M}^+T_\Theta G^-$. Hence $Bv \cap Bv' = 0$ and $vB \cap v'B = 0$ for any two linearly independent elements v and v' of $T'N^+$. Therefore the multiplicaton map induces injections from $B \otimes \mathcal{C}[T'] \otimes N^+$ and $N^+ \otimes \mathcal{C}[T'] \otimes B$ into U.

Recall the conjugate linear antiautomorphism $\kappa = \kappa_B$ of U which restricts to an antiautomorphism of B defined in Section 1 (see (1.22)). It follows from (1.22) that $\kappa(G^-) = U^+$ and $\kappa(U^+) = G^-$. It further follows from (1.22) and (1.24) that $\kappa(N^-) = N^+$. Applying κ to the above yields injections from $B \otimes \mathcal{C}[T'] \otimes N^-$ and $N^- \otimes \mathcal{C}[T'] \otimes B$ into U.

Since $T = T' \times T_\Theta$, Lemma 4.1(i) ensures that $U^+G^- \subseteq N^+T'B$. It follows that $U = U^+G^-T \subseteq N^+T'B$. Hence $U = N^+T'B$ and isomorphism (i) now follows. Assertion (iii) follows similarly from Lemma 4.1(iii). Isomorphism (ii) now follows from (iii) and (iv) from (i) by applying the antiautomorphism κ. $\square$

Recall the definition of the groups $\check{\mathcal{A}}$ and $\check{T}_\Theta$ given in Section 3. Consider for the moment $a \in \check{U}^B$. Note that $tat^{-1} = a$ for all $t \in T_\Theta$. It follow that a is a sum of weight vectors each of whose weight β satisfies $\Theta(\beta) = -\beta$. Thus the invariant ring $\check{U}^B$ is equal to $\check{U}^{B\check{T}_\Theta}$.

We are now ready to define the quantum Harish-Chandra projection map with respect to a subalgebra B in $\mathcal{B}$. Theorem 4.2 and the decomposiition in (3.5) imply the following inclusion for each $B \in \mathcal{B}$

$$\check{U} \subseteq ((B\check{T}_\Theta)_+\check{U}\check{\mathcal{A}} + N_+^+\check{\mathcal{A}}) \oplus \mathcal{C}[\check{\mathcal{A}}]. \tag{4.8}$$

DEFINITION 4.3. The (quantum) Harish-Chandra map with respect to the symmetric pair $\mathfrak{g}, \mathfrak{g}^\theta$ and subalgebra B in $\mathcal{B}$ is the projection $\mathcal{P}_B$ of $\check{U}$ onto $\mathcal{C}[\check{\mathcal{A}}]$ using the direct sum decomposition (4.8).

It should be noted that the map $\mathcal{P}_B$ is the same as the map $\mathcal{P}_\mathcal{A}$ defined for standard analogs of $U(\mathfrak{g}^\theta)$ where Σ is a reduced root system in [L5, immediately following (4.11)]. Note further that $\mathcal{P}_B$ is a linear map for each $B \in \mathcal{B}$. However,

$\mathcal{P}_B$ is not an algebra homomorphism on $\check{U}$. The next result shows that $\mathcal{P}_B$ is an algebra homomorphism upon restriction to $\check{U}^B$.

THEOREM 4.4. *For all $B \in \mathcal{B}$ the restriction of $\mathcal{P}_B$ to $\check{U}^B$ is an algebra homomorphism.*

PROOF. Suppose that a and a' are in $\check{U}^B$. We have

$$aa' \in a((B\check{T}_\Theta)_+\check{U} + N_+^+\check{A}) + a\mathcal{P}_B(a').$$

Since $a \in \check{U}^B$, it follows that $a(B\check{T}_\Theta)_+\check{U} \subseteq (B\check{T}_\Theta)_+\check{U}$. Using (4.8) we have

$$aN_+^+\check{A} \subseteq \check{U}N_+^+\check{A} = ((B\check{T}_\Theta)_+\check{U} + N^+\check{A})N_+^+\check{A} \subseteq (B\check{T}_\Theta)_+\check{U} + N_+^+\check{A}.$$

Hence

$$\begin{aligned} aa' &\in (B\check{T}_\Theta)_+\check{U} + (aN_+^+\check{A}) + a\mathcal{P}_B(a') \\ &\subseteq (B\check{T}_\Theta)_+\check{U} + N_+^+\check{A} + \mathcal{P}_B(a)\mathcal{P}_B(a'). \end{aligned}$$

It follows that $\mathcal{P}_B(aa') = \mathcal{P}_B(a)\mathcal{P}_B(a')$. The theorem now follows from the fact that $\mathcal{P}_B$ is a linear map. □

Recall the definition of the subgroup $\mathcal{A}$ of $\check{A}$ given in Section 3. Let G^+ denote the subalgebra of U generated by $x_i t_i^{-1}$ for $1 \leq i \leq n$ and let U^- denote the subalgebra of U generated by y_i for $1 \leq i \leq n$.

LEMMA 4.5. *Suppose that $u \in \check{U}^B$. Then*

$$u \in U^+G^-\mathcal{A} + U^+G^-\mathcal{A}\mathcal{C}[\check{T}_\Theta]_+.$$

PROOF. Recall that $\check{U}^B$ is a subset of $F_r(\check{U})$ (see Theorem 1.2). Recall (1.17) that $F_r(\check{U})$ is a direct sum of $\operatorname{ad}_r U$ submodules of the form $(\operatorname{ad}_r U)\tau(2\gamma)$ with $\gamma \in P^+(2\pi)$. Thus it is sufficient to prove the lemma for $u \in \check{U}^B \cap (\operatorname{ad}_r U)\tau(2\gamma)$ where $\gamma \in P^+(\pi)$. It follows from the formulas for the right adjoint action in (1.14) (or see [L5, discussion preceding Lemma 7.2]), that

$$(4.9) \qquad (\operatorname{ad}_r U)\tau(2\gamma) \subseteq \sum_{\mu \in Q^+(\pi)} G^+U^-\tau(2\gamma - 2\mu).$$

Now $(\operatorname{ad}_r U)\tau(2\gamma)$ can be written as a direct sum of simple $(\operatorname{ad}_r U)$ modules. In particular, u is a sum of $(\operatorname{ad}_r B)$ invariant vectors contained in simple $(\operatorname{ad}_r U)$ submodules of $(\operatorname{ad}_r U)\tau(2\gamma)$. It follows from [L4, Theorem 3.6] that $u = \sum_{\tilde{\lambda} \in P(\Sigma)} w_{2\tilde{\lambda}}$ for some weight vectors $w_{2\tilde{\lambda}}$ in $(\operatorname{ad}_r U)\tau(2\gamma)$ of weight $2\tilde{\lambda}$. Using (4.9), we can write

$$w_{2\tilde{\lambda}} \in \sum_{\mu \in Q^+(\pi)} \sum_{\alpha - \beta = 2\tilde{\lambda}} G_\alpha^+ U_{-\beta}^- \tau(2\gamma - 2\mu)$$

for each $\tilde{\lambda} \in P(\Sigma)$. Now consider α and β in $Q^+(\pi)$ such that $\alpha - \beta = 2\tilde{\lambda}$. Note that $G_\alpha^+ = U_\alpha^+\tau(-\alpha)$ and $U_\beta^- = G^-\tau(-\beta)$. Hence $G_\alpha^+U_\beta^- = U^+G^-\tau(-2\alpha)\tau(\alpha-\beta)$. Since $\alpha - \beta = 2\tilde{\lambda}$, it follows that $\tau(\alpha - \beta) \in \mathcal{A}$. Therefore,

$$u \in \sum_{\eta \in Q(\pi)} U^+G^-\tau(2\eta)\mathcal{A}.$$

Now $\tilde{\eta} \in Q(\Sigma)$ and hence $\tau(2\tilde{\eta}) \in \mathcal{A}$ for all $\eta \in Q(\pi)$. The lemma now follows from the decomposition in (3.5). □

Consider a weight $\eta \in Q^+(\pi)$ such that

$$\eta \in \sum_{\alpha_i \in \pi} \mathbf{N}2\alpha_i + \sum_{\alpha_j \in \mathcal{S}} \mathbf{N}\alpha_j.$$

It follows from [L4, Lemma 3.5] that $\tilde{\eta}/2 \in P(\Sigma)$. Hence $\tau(\tilde{\eta}) \in \mathcal{A}$. Now $\alpha_i + \Theta(\alpha_i) = 0$ whenever $\alpha_i \in \mathcal{S}$. On the other hand, if $\eta \in \sum_{\alpha_i \in \pi} \mathbf{N}2\alpha_i$, then both $\eta/2$ and $\Theta(\eta)/2$ are elements of $Q(\pi)$. Therefore, the assumptions on η force $\tau((\eta + \Theta(\eta))/2)$ to be an element of T. Since $(\eta + \Theta(\eta))/2$ is Θ invariant, we further have that $\tau((\eta+\Theta(\eta))/2) \in T_\Theta$. Thus $\tau(\eta) = \tau(\tilde{\eta})\tau((\eta+\Theta(\eta))/2)$ is an element of $\mathcal{A}T_\Theta$. This fact combined with the previous lemma allows us to refine the codomain of $\mathcal{P}_B$.

THEOREM 4.6. *For all $B \in \mathcal{B}$, the image of $\check{U}^B$ under $\mathcal{P}_B$ is contained in $\mathcal{C}[\mathcal{A}]$.*

PROOF. Let $u \in \check{U}^B$. By Lemma 4.5, there exists $u' \in U^+G^-\mathcal{A}$ such that $\mathcal{P}_B(u) = \mathcal{P}_B(u')$. Note that $\mathcal{A}N_+^+ = N_+^+\mathcal{A}$. It follows from Lemma 4.1(ii) that $u' \in B_+U + \sum_\eta \tau(\eta)N^+$ where η runs over elements in

$$\sum_{\alpha_i \in \pi \setminus \pi_\Theta} \mathbf{N}2\alpha_i + \sum_{\alpha_j \in \mathcal{S}} \alpha_j. \tag{4.10}$$

By the discussion preceding the lemma, we have that $\tau(\eta) \in \mathcal{A}T_\Theta$ for all η in the set given by (4.10). Thus $\mathcal{P}_B(u) \in \mathcal{C}[\mathcal{A}]$. □

CHAPTER 5

Quantum Radial Components

Throughout this section, we assume that B is a fixed coideal subalgebra in $\mathcal{B}$. In this section, we study the map $\mathcal{X}$ on the algebra $\check{U}^B$ which computes the quantum radial components. The map $\mathcal{X}$ is defined in [L5, Section 3] for standard analogs associated to symmetric pairs with reduced root systems. After extending the definition to the general case, we explore connections between $\mathcal{X}$ and the Harish-Chandra projection map $\mathcal{P}_B$. In particular, we show that $\mathcal{X}(\check{U}^B)$ and $\mathcal{P}_B(\check{U}^B)$ are isomorphic as algebras. As in [L5, Theorem 3.4], one has that the quantum radial components are invariant under the action of the restricted Weyl group W_Θ. Using this fact, we determine the possible top degree terms of elements in $\mathcal{P}_B(\check{U}^B)$ with respect to a certain filtration of $\check{U}$. This is a crucial step towards finding the image of $\check{U}^B$ under the quantum Harish-Chandra map $\mathcal{P}_B$.

Before defining the function $\mathcal{X}$, we review some notions from [L5, Section 3] which are necessary to define the codomain of this function. Let $\mathcal{C}[Q(\Sigma)]\mathcal{A}$ denote the ring generated by $\mathcal{C}[Q(\Sigma)]$ and $\mathcal{C}[\mathcal{A}]$ subject to the exchange rule

$$z^\lambda \tau(\mu) = q^{(\lambda,\mu)}\tau(\mu)z^\lambda. \tag{5.1}$$

Note that the ring $\mathcal{C}[P(\Sigma)]$ is both a right $\mathcal{C}[\mathcal{A}]$ module and a right $\mathcal{C}[Q(\Sigma)]$ module where elements of $\mathcal{C}[Q(\Sigma)]$ act by (right) multiplication and

$$z^\lambda * \tau(\mu) = q^{(\lambda,\mu)}z^\lambda \tag{5.2}$$

for all $\lambda \in P(\Sigma)$ and $\tau(\mu) \in \mathcal{A}$. In particular, both $\mathcal{C}[\mathcal{A}]$ and $\mathcal{C}[Q(\Sigma)]$ embed in the right endomorphism ring $\mathrm{End}_r\mathcal{C}[P(\Sigma)]$. The algebra $\mathcal{C}[Q(\Sigma)]\mathcal{A}$ can be identified with the subalgebra of $\mathrm{End}_r\mathcal{C}[P(\Sigma)]$ generated by $\mathcal{C}[Q(\Sigma)]$ and $\mathcal{C}[\mathcal{A}]$. (Indeed, this is how $\mathcal{C}[Q(\Sigma)]\mathcal{A}$ is defined in [L5, Section 3].) One can localize by the nonzero elements of $\mathcal{C}[Q(\Sigma)]$ to form a new ring $\mathcal{C}(Q(\Sigma))\mathcal{A}$ generated by $\mathcal{A}$ and the quotient field $\mathcal{C}(Q(\Sigma))$ of $\mathcal{C}[Q(\Sigma)]$.

Note that $\mathcal{C}[\mathcal{A}]$ is a left $\mathcal{C}[Q(\Sigma)]\mathcal{A}$ module where elements of $\mathcal{A}$ act via left multiplication and

$$z^\lambda \cdot \tau(\mu) = q^{(\lambda,\mu)}\tau(\mu)$$

for all $\lambda \in Q(\Sigma)$ and $\tau(\mu) \in \mathcal{A}$. Note further that this action is compatible with the action given in (5.2). In particular,

$$a'(b \cdot a) = (a' * b)(a) \tag{5.3}$$

for all $a' \in \mathcal{C}[P(\Sigma)]$, $b \in \mathcal{C}[Q(\Sigma)]\mathcal{A}$, and $a \in \mathcal{C}[\mathcal{A}]$. It should also be noted that the above actions can all be extended to the setting where $\mathcal{A}$ is replaced by $\check{\mathcal{A}}$.

Now $g \cdot \tau(\mu)$ is just a scalar multiple of $\tau(\mu)$ for $\tau(\mu) \in \mathcal{A}$ and $g \in \mathcal{C}[Q(\Sigma)]$. Given an element $f \in \mathcal{C}[Q(\Sigma)]\mathcal{A}$ and $g \in \mathcal{C}[Q(\Sigma)]$, set

$$(fg^{-1}) \cdot \tau(\mu) = (f \cdot \tau(\mu))((g \cdot \tau(\mu))\tau(\mu)^{-1})^{-1}$$

for all $\tau(\mu) \in \mathcal{A}$ such that $(g \cdot \tau(\mu))$ is nonzero.

Set $\mathcal{A}_{\geq} = \{\tau(\mu) | \mu \in Q^+(\Sigma) \cap P(2\Sigma)\}$. Alternatively, we may view $\mathcal{A}_{\geq}$ as the submonoid of $\mathcal{A}$ generated by $\tau(\tilde{\alpha}_j)$ for $\alpha_j \in \mathcal{S}$ and elements in the set $\{\tau(2\tilde{\alpha}) | \tilde{\alpha} \in \pi \setminus \pi_\Theta\}$. Set $U_{\geq} = U^+G^-\mathcal{A}_{\geq}$. Set $B' = \chi(B)$ where χ is the Hopf algebra automorphism of U defined by (1.27).

LEMMA 5.1. *For each $\tau(\gamma) \in \check{\mathcal{A}}$ and $X \in U^+G^-\tau(\gamma)$, there exists $p_X \in \mathcal{C}(Q(\Sigma))\mathcal{A}_{\geq}\tau(\gamma)$ such that*

$$X\tau(\eta) - (p_X \cdot \tau(\eta)) \in B_+U_{\geq}\tau(\gamma+\eta) + U_{\geq}\tau(\gamma+\eta)B'_+$$

for all $\tau(\eta) \in \check{\mathcal{A}}$ such that $p_X \cdot \tau(\eta)$ is defined.

PROOF. This is just a slight generalization of [L5, Lemma 3.1] to all quantum symmetric pairs. We follow the proof in [L5]. The first step is the construction of linear maps from the restricted weight spaces of N^+ to N^- and from the restricted weight spaces of N^- to N^+ along the lines of [L5, Theorem 2.3]. In particular, by Lemma 4.1 and the discussion preceding Theorem 4.6, we see that

$$N^-_{-\beta,r} \subseteq N^+_{\beta,r} + \sum_{\tilde{\gamma}<\tilde{\beta}} N^+_{\gamma,r}\mathcal{A}_{\geq} + U_{\geq}B'_+ \tag{5.4}$$

for all $\beta \in Q^+(\pi)$. For each $\tilde{\beta} \in Q^+(\Sigma)$, we can define a linear map (using (5.4))

$$P_{\tilde{\beta}} : N^-_{-\beta,r} \mapsto N^+_{\beta,r}$$

such that

$$Y - P_{\tilde{\beta}}(Y) \in \sum_{\tilde{\gamma}<\tilde{\beta}} N^+_{\gamma,r}\mathcal{A}_{\geq} + U_{\geq}B'_+$$

for all $Y \in N^-_{-\beta,r}$.

Now Theorem 4.2(iii) implies that

$$N^- \cap (U(\mathcal{A}_{\geq})_+ + UB'_+)$$

is the empty set. On the other hand, N^- can be written as a direct sum of its restricted weight spaces $N^-_{-\gamma,r}$. It follows that $P_{\tilde{\beta}}$ is injective. Since $N^+_{\beta,r}$ and $N^-_{-\beta,r}$ are finite-dimensional vector spaces of the same dimension (see [L5, proof of Theorem 2.3]), we further have that $P_{\tilde{\beta}}$ is an isomorphism. A similar argument using Lemma 4.1 and Theorem 4.2(ii) yield a vector space isomorphism

$$R_{\tilde{\beta}} : N^+_{\beta,r} \mapsto N^-_{-\beta,r}$$

such that

$$X - R_{\tilde{\beta}}(X) \in \sum_{\tilde{\gamma}<\tilde{\beta}} N^-_{-\gamma,r}\mathcal{A}_{\geq} + B_+U_{\geq}$$

for all $X \in N^+_{\gamma,r}$. It follows that the composition $P_{\tilde{\beta}} \circ R_{\tilde{\beta}}$ is an isomorphism of $N^+_{\beta,r}$ onto itself. Choose a basis $X_i, 1 \leq i \leq m$ for $N^+_{\beta,r}$ so that $P_{\tilde{\beta}} \circ R_{\tilde{\beta}}$ corresponds to an upper triangular matrix with diagonal entries c_{ii}. The fact that $P_{\tilde{\beta}} \circ R_{\tilde{\beta}}$ is invertible ensures that the c_{ii} are nonzero.

By Lemma 4.1 and the discussion preceding Theorem 4.6, we may reduce to the case when $X \in N^+\mathcal{A}_{\geq}\tau(\gamma)$. Note that if X and X' both satisfy the conclusions of the lemma, then so does $X + X'$. Hence we may assume that $X \in N^+_{\beta,r}\tau(\gamma)$ for some $\beta \in Q^+(\pi)$ and $\tau(\gamma) \in \check{\mathcal{A}}$. We have

$$X_i\tau(\gamma)\tau(\eta) \in R_{\tilde{\beta}}(X_i)\tau(\gamma+\eta) + \sum_{\tilde{\xi}<\tilde{\beta}} N^-_{-\xi,r}\mathcal{A}_{\geq}\tau(\gamma+\eta) + B_+U_{\geq}\tau(\gamma+\eta)$$

for all $\tau(\eta) \in \check{\mathcal{A}}$. Note that $N^-_{-\xi,r}\tau(\zeta) = \tau(\zeta)N^-_{-\xi,r}$ for all $\tau(\zeta) \in \check{\mathcal{A}}$. Hence (5.4) combined with the fact that the restricted weight of $R_{\tilde{\beta}}(X_i)$ is $-\tilde{\beta}$ yields

$$X_i\tau(\gamma)\tau(\eta) \in q^{(\gamma+\eta,\tilde{\beta})}\tau(\gamma+\eta)R_{\tilde{\beta}}(X_i) + \sum_{\tilde{\xi}<\tilde{\beta}} N^+_{\xi,r}\mathcal{A}_{\geq}\tau(\gamma+\eta) + U_{\geq}\tau(\gamma+\eta)B'_+.$$

Applying the map $P_{\tilde{\beta}}$ to $R_{\tilde{\beta}}(X_i)$ now gives us

$$\begin{aligned} X_i\tau(\gamma)\tau(\eta) &\in q^{(\gamma+\eta,\tilde{\beta})}\tau(\gamma+\eta)P_{\tilde{\beta}}\circ R_{\tilde{\beta}}(X_i) + \sum_{\tilde{\xi}<\tilde{\beta}} N^+_{\xi,r}\mathcal{A}_{\geq}\tau(\gamma+\eta) \\ &\quad + B_+U_{\geq}\tau(\gamma+\eta) + U_{\geq}\tau(\gamma+\eta)B'_+ \\ &= q^{(2\gamma+2\eta,\tilde{\beta})}P_{\tilde{\beta}}\circ R_{\tilde{\beta}}(X_i)\tau(\gamma+\eta) + \sum_{\tilde{\xi}<\tilde{\beta}} N^+_{\xi,r}\mathcal{A}x_{\geq}\tau(\gamma+\eta) \\ &\quad + B_+U_{\geq}\tau(\gamma+\eta) + U_{\geq}\tau(\gamma+\eta)B'_+. \end{aligned} \tag{5.5}$$

Note that the choice of $X_i, 1 \leq i \leq m$ ensures that

$$P_{\tilde{\beta}}\circ R_{\tilde{\beta}}(X_i) \in c_{ii}X_i + \sum_{j<i}\mathcal{C}X_j.$$

Hence (5.5) implies that

$$\begin{aligned} (1 - c_{ii}q^{(2\gamma+2\eta,\tilde{\beta})})X_i\tau(\gamma)\tau(\eta) &\in \sum_{j<i}\mathcal{C}X_j\tau(\gamma+\eta) + \sum_{\tilde{\xi}<\tilde{\beta}} N^+_{\xi,r}\mathcal{A}_{\geq}\tau(\gamma+\eta) \\ &\quad + B_+U_{\geq}\tau(\gamma+\eta) + U_{\geq}\tau(\gamma+\eta)B'_+. \end{aligned}$$

Using induction on i and ξ, we see that there exists $p \in \mathcal{C}(Q(\Sigma))\mathcal{A}_{\geq}\tau(\gamma)$ such that

$$(1 - c_{ii}q^{(2\gamma+2\eta,\tilde{\beta})})X_i\tau(\gamma)\tau(\eta) + p\cdot\tau(\eta) \in B_+U_{\geq}\tau(\gamma+\eta) + U_{\geq}\tau(\gamma+\eta)B'_+.$$

Set $p_X = -(1 - c_{ii}q^{(2\gamma,\tilde{\beta})}z^{2\tilde{\beta}})^{-1}p$. □

Recall the isomorphism between the space ${}_{B'}\mathcal{H}_B$ of right B and left B' invariants inside $R_q[G]$ and the ring $\mathcal{C}[P(2\Sigma)]^{W_\Theta}$ given in Theorem 1.3. The formula for the evaluation of the function z^λ at the element $\tau(\gamma)$ of $\check{U}^0$ given in (1.18) leads to a vector space pairing between $\mathcal{C}[P(2\Sigma)]$ and $\mathcal{C}[\mathcal{A}]$. Moreover, this pairing restricts to a nondegenerate pairing between $\mathcal{C}[P(2\Sigma)]^{W_\Theta}$ and $\mathcal{C}[\mathcal{A}]^{W_\Theta}$. Thus if $a \in \mathcal{C}[\mathcal{A}]$ and $\sum_{w\in W_\Theta} wa$ is nonzero, then there exists $g \in {}_{B'}\mathcal{H}_B$ such that $g(a) \neq 0$.

Now suppose that $X \in U^+G^-\tau(\gamma)$ and let p_X be an element in the ring $\mathcal{C}(Q(\Sigma))\mathcal{A}_{\geq}\tau(\gamma)$ that satisfies the conclusion of Lemma 5.1. We show that p_X is unique. Indeed suppose that p'_X is another element in $\mathcal{C}(Q(\Sigma))\mathcal{A}_{\geq}\tau(\gamma)$ that satisfies the conclusion of Lemma 5.1. It follows that

$$(p_X - p'_X)\cdot\tau(\eta) \in B_+U\tau(\gamma+\eta) + U\tau(\gamma+\eta)B'_+$$

for all η such that $(p_X - p'_X)\cdot\tau(\eta)$ is defined. Note further that $g(x) = 0$ for all $x \in B_+U\tau(\gamma+\eta) + U\tau(\gamma+\eta)B'_+$ and $g \in_{B'} \mathcal{H}_B$. Thus, the equality $p_X = p'_X$ follows from the next lemma.

LEMMA 5.2. *Suppose that f is a nonzero element of $\mathcal{C}(Q(\Sigma))\mathcal{A}_{\geq}\tau(\gamma)$ for some $\tau(\gamma) \in \check{\mathcal{A}}$. Then there exists $\tau(\eta) \in \check{\mathcal{A}}$ and $g \in {}_{B'}\mathcal{H}_B$ such that $f\cdot\tau(\eta)$ is defined and $g(f\cdot\tau(\eta)) \neq 0$.*

PROOF. We can write f as a sum of the form $\sum_{i=1}^{n} f_i\tau(\beta_i)$ where each $\tau(\beta_i) \in \mathcal{A}\tau(\gamma)$ and each f_i is a quotient of Laurent polynomials in $\mathcal{C}[Q(\Sigma)]$. It follows that there exists $\tau(\eta) \in \check{\mathcal{A}}$ such that $\tau(\gamma+\eta) \in \mathcal{A}$ and $f_i \cdot \tau(\eta)$ is defined and nonzero for $i = 1, \ldots, n$. In other words, $(\sum_i f_i\tau(\beta_i)) \cdot \tau(\eta)$ is a nonzero element of $\mathcal{C}[\mathcal{A}]$. Note that $(\sum_i f_i\tau(\beta_i)) \cdot \tau(\eta)$ is just a linear combination of the $\tau(\beta_i+\eta)$ for $1 \leq i \leq n$. We can further choose η large enough so that $\beta_i + \eta$ is a dominant integral restricted weight for each i. In particular, we can assume that η has been chosen so that $(\sum_i f_i\tau(\beta_i)) \cdot \tau(\eta)$ is a nonzero linear combination of terms of the form $\tau(\beta_i+\eta)$ where each $\beta_i + \eta$ is a dominant restricted weight. It follows that $\sum_{w\in W_\Theta} w(\sum_i f_i\tau(\beta_i) \cdot \tau(\eta))$ is nonzero. By the discussion preceding the lemma, there exists $g \in_{B'} \mathcal{H}_B$ such that $g((\sum_i f_i\tau(\beta_i)) \cdot \tau(\eta)) \neq 0$. □

One of the important properties of the zonal spherical functions $g_{2\lambda}$, $\lambda \in P^+(\Sigma)$, is that they are joint eigenfunctions with respect to the right action of elements of $\check{U}^B$ (Theorem 1.4). In the next lemma we see that the eigenvalues can be determined using the quantum Harish-Chandra map $\mathcal{P}_B$. The proof is based on the discussion following [L5, Lemma 3.5].

LEMMA 5.3. *Let $\lambda \in P^+(\Sigma)$. Then*

$$g_{2\lambda} \cdot c = z^{2\lambda}(\mathcal{P}_B(c))g_{2\lambda} \tag{5.6}$$

for all $c \in \check{U}^B$.

PROOF. Write $\xi^*_{2\lambda}$ for a nonzero vector in $(L(2\lambda)^*)^B$ and $\xi_{2\lambda}$ for a nonzero vector in $L(2\lambda)^B$. Note that $g_{2\lambda}$ is just a scalar multiple of $\xi_{2\lambda} \otimes \xi^*_{2\lambda}$. Let $v^*_{2\lambda}$ denote the lowest weight generating vector of $L(2\lambda)^*$. By [L4, Lemma 3.3], rescaling if necessary, we have

$$\xi^*_{2\lambda} \in v^*_{2\lambda} + v^*_{2\lambda}N^+_+. \tag{5.7}$$

Given $c \in \check{U}^B$, we can write $c = \mathcal{P}_B(c)+m+b$ for some $m \in N^+_+\mathcal{A}$ and $b \in (B\check{T}_\Theta)_+\check{U}$. Hence

$$\xi^*_{2\lambda}c \in v^*_{2\lambda}\mathcal{P}_B(c) + v^*_{2\lambda}N^+_+ = z^{2\lambda}(\mathcal{P}_B(c))v^*_{2\lambda} + v^*_\lambda N^+_+.$$

By Theorem 1.1, $\xi^*_{2\lambda}c$ is just a scalar multiple of $\xi^*_{2\lambda}$. By (5.7) and the above expression we see that $\xi^*_{2\lambda}c = z^{2\lambda}(\mathcal{P}_B(c))\xi^*_{2\lambda}$. The lemma now follows. □

Recall that $\varphi_{2\lambda}$ is the image of the zonal spherical function $g_{2\lambda}$ under the restriction map Υ (see Section 1). Recall further that the set $\{g_{2\lambda} | \; \lambda \in P^+(\Sigma)\}$ is a basis for ${}_{B'}\mathcal{H}_B$ and $\{\varphi_{2\lambda} | \; \lambda \in P^+(\Sigma)\}$ is a basis for $\mathcal{C}[P(2\Sigma)]^{W_\Theta}$.

The next two theorems generalize [L5, Theorem 3.6].

THEOREM 5.4. *There is a linear map $\mathcal{X}$ from $\check{U}^B$ to the ring $\mathcal{C}(Q(\Sigma))\mathcal{A}$ such that*

$$\varphi_{2\lambda} * \mathcal{X}(c) = z^{2\lambda}(\mathcal{P}_B(c))\varphi_{2\lambda} \tag{5.8}$$

and

$$g_{2\lambda}(c\tau(\beta)) = (\varphi_{2\lambda} * \mathcal{X}(c))(\tau(\beta)) \tag{5.9}$$

for all $c \in \check{U}^B$, $\lambda \in P^+(\Sigma)$, and $\tau(\beta) \in \mathcal{A}$.

PROOF. Using Lemma 5.1 and Lemma 5.2, we can define a linear map

$$\mathcal{X} : U\mathcal{A} \to \mathcal{C}(Q(\Sigma))\mathcal{A}$$

such that

$$g(\mathcal{X}(x) \cdot \tau(\eta)) = g(x\tau(\eta)) \tag{5.10}$$

for all $x \in U\mathcal{A}$, $g \in {}_{B'}\mathcal{H}_B$, and $\tau(\eta) \in \mathcal{A}$ with $\mathcal{X}(x) \cdot \tau(\eta)$ defined. Note that $g(y) = 0$ for all $y \in \mathcal{C}[\check{T}_\Theta]_+U + U\mathcal{C}[\check{T}_\Theta]_+$. Thus the map $\mathcal{X}$ extends to a linear map on $U\mathcal{A}\check{T}_\Theta$ such that (5.10) still holds. In particular, by Lemma 4.5, $\mathcal{X}$ restricts to a linear map on $\check{U}^B$. Fix $c \in \check{U}^B$ and $\lambda \in P^+(\Sigma)$. Since $\varphi_{2\lambda}$ is just the image of $g_{2\lambda}$ under the restriction map Υ, it follows that $\varphi_{2\lambda}(\tau(\zeta)) = g_{2\lambda}(\tau(\zeta))$ for all $\tau(\zeta) \in \mathcal{A}$. Combining Lemma 5.3 and (5.10) thus yields

$$\varphi_{2\lambda}(\mathcal{X}(c) \cdot \tau(\eta)) = g_{2\lambda}(c\tau(\eta)) = z^{2\lambda}(\mathcal{P}_B(c))\varphi_{2\lambda}(\tau(\eta))$$

for all $\tau(\eta) \in \mathcal{A}$ with $\mathcal{X}(c) \cdot \tau(\eta)$ defined. Moreover, the compatibility of actions given in (5.3) yields

$$(\varphi_{2\lambda} * \mathcal{X}(c) - z^{2\lambda}(\mathcal{P}_B(c))\varphi_{2\lambda})(\tau(\eta)) = 0 \tag{5.11}$$

for all $\tau(\eta) \in \mathcal{A}$ with $\mathcal{X}(c) \cdot \tau(\eta)$ defined.

Note that $\varphi_{2\lambda}$ is a Laurent polynomial in $\mathcal{C}[P(\Sigma)]$ and so $\varphi_{2\lambda} \cdot \tau(\eta)$ is defined for all $\tau(\eta) \in \mathcal{A}$. On the other hand, we can find a nonzero element $p \in \mathcal{C}[Q(\Sigma)]$ such that $p\mathcal{X}(c) \in \mathcal{C}[Q(\Sigma)]\mathcal{A}$. In particular, if $\mathcal{X}(c) \cdot \tau(\eta)$ is undefined then $p \cdot \tau(\eta) = 0$. This implies that (5.11) holds for all η in a Zariski dense subset of $\sum_{\alpha_i \in \pi^*} \mathbf{Q}\tilde{\alpha}_i$. Assertions (5.8) and (5.9) of the theorem now follow. □

Suppose that $c \in \check{U}^B$. As in the classical theory, we call $\mathcal{X}(c)$ the *radial component* of c. Moreover, we often refer to the map $\mathcal{X}$ as the *radial component map*. Technically, $\mathcal{X}$ depends on the choice of B. This choice will be understood from context.

The restricted Weyl group W_Θ acts on $\mathcal{C}[Q(\Sigma)]\mathcal{A}$ where $w\tau(\beta) = \tau(w\beta)$ and $w(z^\mu) = z^{w\mu}$ for all $w \in W_\Theta$, $\tau(\beta) \in \mathcal{A}$ and $\mu \in Q(\Sigma)$. This action of W_Θ on $\mathcal{C}[Q(\Sigma)]\mathcal{A}$ extends to an action of W_Θ on $\mathcal{C}(Q(\Sigma))\mathcal{A}$. The next result shows that the image of $\check{U}^B$ under the radial component map is W_Θ invariant.

THEOREM 5.5. *The map $\mathcal{X}$ is an algebra homomorphism from $\check{U}^B$ to the W_Θ invariant ring $(\mathcal{C}(Q(\Sigma))\mathcal{A})^{W_\Theta}$.*

PROOF. By Theorem 5.4, we know that $\mathcal{X}$ is a linear map. Suppose that c, c' are two elements of $\check{U}^B$. By (5.8) we have

$$(\varphi_{2\lambda} * \mathcal{X}(cc')) = z^{2\lambda}(\mathcal{P}_B(cc'))\varphi_{2\lambda}$$

for all $\lambda \in P^+(\Sigma)$. By Theorem 4.4, $\mathcal{P}_B$ is an algebra homomorphism when restricted to $\check{U}^B$. Hence

$$(\varphi_{2\lambda} * \mathcal{X}(cc')) = z^{2\lambda}(\mathcal{P}_B(c))z^{2\lambda}(\mathcal{P}_B(c'))\varphi_{2\lambda} = (\varphi_{2\lambda} * \mathcal{X}(c')) * \mathcal{X}(c)$$

for each choice of λ. The fact that $\mathcal{P}_B(\check{U}^B)$ is a subring of the commutative ring $\mathcal{C}[\mathcal{A}]$ yields that $\mathcal{X}$ is an algebra homomorphism upon restriction to $\check{U}^B$.

It remains to check that the image of $\check{U}^B$ under $\mathcal{X}$ is W_Θ invariant. Recall (Theorem 1.3 and subsequent discussion) that the set $\{\varphi_{2\lambda} | \lambda \in P^+(\Sigma)\}$ is a basis for $\mathcal{C}[P(2\Sigma)]^{W_\Theta}$. It follows that

$$\varphi_{2\lambda} * (\mathcal{X}(c) - w(\mathcal{X}(c))) = 0$$

for all $w \in W_\Theta$ and $\lambda \in P^+(\Sigma)$. Suppose that $\mathcal{X}(c) - w(\mathcal{X}(c))$ is nonzero. By Lemma 5.2, we can find $\tau(\eta) \in \mathcal{A}$ such that $g((\mathcal{X}(c) - w(\mathcal{X}(c)) \cdot \tau(\eta))$ is nonzero for some $g \in {}_{B'}\mathcal{H}_B$. Hence, there exists $\lambda \in P^+(\Sigma)$ such that

$$g_{2\lambda}((\mathcal{X}(c) - w(\mathcal{X}(c)) \cdot \tau(\eta)) \neq 0.$$

This contradiction forces $\mathcal{X}(c)$ to be W_Θ invariant. □

Note that the fact that $\mathcal{X}$ is an algebra homomorphism is established in the above proof using (5.8) and properties of the Harish-Chandra map $\mathcal{P}_B$. The next theorem makes more precise the relationship between the two maps $\mathcal{X}$ and $\mathcal{P}_B$.

THEOREM 5.6. *The map $\mathcal{X}(c) \to \mathcal{P}_B(c)$ defines an algebra isomorphism from $\mathcal{X}(\check{U}^B)$ onto $\mathcal{P}_B(\check{U}^B)$. Moreover, the kernel of both $\mathcal{X}$ and $\mathcal{P}_B$ upon restriction to $\check{U}^B$ is equal to $\check{U}^B \cap (B\check{T}_\Theta)_+\check{U}\check{\mathcal{A}}$.*

PROOF. Consider an element $c \in \check{U}^B$. Note that $\mathcal{P}_B(c) = 0$ if and only if $z^{2\lambda}(\mathcal{P}_B(c)) = 0$ for all $\lambda \in P^+(\Sigma)$. On the other hand, if $\mathcal{X}(c) \neq 0$ then Lemma 5.2 ensures that $g_{2\lambda} * \mathcal{X}(c)$ is nonzero for some $\lambda \in P^+(\Sigma)$. It follows that $\mathcal{P}_B(c) = 0$ if and only if $\mathcal{X}(c) = 0$. Therefore the map from $\mathcal{X}(\check{U}^B)$ to $\mathcal{P}_B(\check{U}^B)$ is well defined and bijective. Assertion (5.8) of Theorem 5.4 further implies that this map is an algebra homomorphism. This establishes the first part of the theorem.

Suppose that $c \in \check{U}^B$ and $\mathcal{P}_B(c) = 0$. Using the definition of $\mathcal{P}_B$ (Definition 4.3), we can write $c = n + b$ where $b \in (B\check{T}_\Theta)_+\check{U}\check{\mathcal{A}}$ and $n \in N_+^+\check{\mathcal{A}}$. Hence it is sufficient to show that $n = 0$. Assume otherwise. We can express n as a linear combination of weight vectors $\sum_\gamma a_\gamma n_\gamma$ where each $n_\gamma \in N_+^+$ and each $a_\gamma \in \mathcal{C}[\check{\mathcal{A}}]$. Let γ_0 be a minimal element in the set $\{\gamma|\ a_\gamma \neq 0\}$ with respect to the standard ordering on $Q^+(\pi)$. Since a_{γ_0} is a nonzero element of $\mathcal{C}[\check{\mathcal{A}}]$, we can choose $\lambda \in P^+(\Sigma)$ such that $z^{2\lambda}(a_{\gamma_0}) \neq 0$. Let $\xi_{2\lambda}^*$ be a nonzero vector in $(L(2\lambda)^*)^B$ and recall the expression for $\xi_{2\lambda}^*$ given in (5.7). We have

$$\begin{aligned} \xi_{2\lambda}^* c = \xi_{2\lambda}^* n &\in v_{2\lambda}^* n + v_{2\lambda}^* N_+^+ n \\ &\subseteq z^{2\lambda}(a_{\gamma_0}) v_{2\lambda}^* n_{\gamma_0} + \sum_{\gamma \not< \gamma_0} v_{2\lambda}^* N_\gamma^+. \end{aligned}$$

In particular, when $\xi_{2\lambda}^*$ is written as a sum of weight vectors, the vector with the lowest weight is a nonzero scalar multiple of $v_{2\lambda}^* n_{\gamma_0}$. However, by Theorem 1.1, we know that $\xi_{2\lambda}^* c$ is just a scalar multiple of $\xi_{2\lambda}^*$. In other words, by (5.7), the lowest weight term of $\xi_{2\lambda}^* c$ is a nonzero scalar multiple of $v_{2\lambda}^*$. This contradiction forces $n = 0$. □

In order to gain more information about the connection between the two maps $\mathcal{X}$ and $\mathcal{P}_B$, it is useful to write radial components as formal power series in the z^μ with coefficients from $\mathcal{C}[\mathcal{A}]$ along the lines of [L5, Section 3]. In particular, let $\mathcal{C}((Q(\Sigma)))$ denote the formal Laurent series ring $\mathcal{C}((z^{-\tilde{\alpha}_i}|\alpha_i \in \pi^*))$. As a set, $\mathcal{C}((Q(\Sigma)))$ consists of finite linear combinations of possibly infinite sums of the form $\sum_{\gamma \leq \beta} a_\gamma z^\gamma$ where γ and β are elements of $Q(\Sigma)$ and each $a_\gamma \in \mathcal{C}$. The embedding of $\mathcal{C}[\mathcal{A}]$ inside $\text{End}_r\mathcal{C}[Q(\Sigma)]$ extends to an embedding of $\mathcal{C}[\mathcal{A}]$ inside $\text{End}_r\mathcal{C}((Q(\Sigma)))$. Let $\mathcal{C}((Q(\Sigma)))\mathcal{A}$ denote the subring of $\text{End}_r\mathcal{C}((Q(\Sigma)))$ generated by $\mathcal{C}((Q(\Sigma)))$ and $\mathcal{A}$. Note that the quotient ring $\mathcal{C}(Q(\Sigma))$ can be identified with a subring of $\mathcal{C}((Q(\Sigma)))$ in a standard way. Thus we may view $\mathcal{C}(Q(\Sigma))\mathcal{A}$ as a subring of $\mathcal{C}((Q(\Sigma)))\mathcal{A}$. Note that elements of $\mathcal{C}((Q(\Sigma)))\mathcal{A}$ are just finite linear combinations of elements of $\mathcal{A}$ with coefficients in $\mathcal{C}((Q(\Sigma)))$. These elements can also be written as finite linear combinations of possibly infinite sums of the form $\sum_{\gamma\leq\beta} a_\gamma z^\gamma$ where γ and β are elements in $Q(\Sigma)$ and the a_γ are Laurent polynomials in $\mathcal{C}[\mathcal{A}]$. (Note, however, that not all elements of this latter form are contained in $\mathcal{C}((Q(\Sigma)))\mathcal{A}$.)

Let $\mathcal{C}[[Q^-(\Sigma)]]$ denote the power series subring of $\mathcal{C}((Q(\Sigma)))$ consisting of possibly infinite sums of the form $\sum_{\gamma\in Q^+(\Sigma)} a_{-\gamma}z^{-\gamma}$ for $a_{-\gamma}\in\mathcal{C}$. Let $\mathcal{C}[[Q^-(\Sigma)]]_+$ denote the augmentation ideal of $\mathcal{C}[[Q^-(\Sigma)]]$. In partiuclar,

$$\mathcal{C}[[Q^-(\Sigma)]]_+ = \sum_{\alpha_i\in\pi^*} z^{-\tilde{\alpha}_i}\mathcal{C}[[Q^-(\Sigma)]]$$

The next lemma provides another connection between the radial components and the Harish-Chandra projections of B invariant elements.

By [L4, Lemma 4.1], we have

$$\varphi_{2\lambda}\in z^{2\lambda}+\sum_{\gamma<2\lambda}\mathcal{C}z^\gamma \tag{5.12}$$

where each γ is in $P(\Sigma)$.

LEMMA 5.7. *For all* $c\in\check{U}^B$,

$$\mathcal{X}(c)\in\mathcal{P}_B(c)+\mathcal{C}[[Q^-(\Sigma)]]_+\mathcal{A}.$$

PROOF. Suppose $c\in\check{U}^B$. Note that $\mathcal{C}[[Q^-(\Sigma)]]\mathcal{A}=\mathcal{A}\mathcal{C}[[Q^-(\Sigma)]]$. In particular, we can write

$$\mathcal{X}(c)=\sum_{i=1}^m\sum_{\gamma\leq\beta_i} b_\gamma z^\gamma$$

where γ and the β_i, for $1\leq i\leq m$, are elements of $Q(\Sigma)$ and the b_γ are Laurent polynomials in $\mathcal{C}[\mathcal{A}]$. Moreover, we may assume that the β_i are not comparable via the standard partial ordering defined by "$<$ " on the weight lattice of restricted weights $P(\Sigma)$.

Since each b_{β_i} is a Laurent polynomial in $\mathcal{C}[\mathcal{A}]$, it follows that there exists $\lambda\in P^+(\Sigma)$ such that $z^{2\lambda}(b_{\beta_i})\neq 0$ for each i. Using (5.12), we have that

$$\varphi_{2\lambda}*\mathcal{X}(c)=\varphi_{2\lambda}*\sum_{i=1}^m\sum_{\gamma\leq\beta_i} b_\gamma z^\gamma\in\sum_{i=1}^m(z^{2\lambda}(b_{\beta_i}))z^{2\lambda+\beta_i}+\sum_{\gamma<\beta_i}\mathcal{C}z^{2\lambda+\gamma}.$$

By Theorem 5.4, $\varphi_{2\lambda}$ is an eigenvector for the action of $\mathcal{X}(c)$ with eigenvalue equal to $z^{2\lambda}(\mathcal{P}_B(c))$ for all $\lambda\in P^+(\Sigma)$. This forces $m=1$, $\beta_1=0$, and $b_0=\mathcal{P}_B(c)$. □

Given an element $\gamma=\sum_{\alpha_i\in\pi^*} m_i\tilde{\alpha}_i$ in $Q(\Sigma)$, we define the restricted height of γ, denoted by $\mathrm{ht}_r\gamma$, by

$$\mathrm{ht}_r\gamma=\sum_{\alpha_i\in\pi^*} m_i. \tag{5.13}$$

Note that elements of $\mathcal{C}[[Q^-(\Sigma)]]\mathcal{A}$ can be written as possibly infinite linear combinations of elements in $\mathcal{A}$ with coefficients in $\mathcal{C}[[Q^-(\Sigma)]]$. The next lemma shows that given $c\in\check{U}^B$, the terms of $\mathcal{A}$ which appear in such an expansion of $\mathcal{X}(c)$ are exactly the same as those terms that appear in $\mathcal{P}_B(c)$.

LEMMA 5.8. *Let $c \in \check{U}^B$ and write $\mathcal{P}_B(c) = \sum_{i=1}^n a_{\beta_i}\tau(\beta_i)$ where each a_{β_i} is a nonzero scalar. Then there exist nonzero elements $f_{\beta_i} \in \mathcal{C}(Q(\Sigma))$ such that $\mathcal{X}(c) = \sum_{i=1}^n f_{\beta_i}\tau(\beta_i)$.*

PROOF. By Lemma 5.7, we may write

$$\mathcal{X}(c) = \sum_{\gamma \geq 0} c_\gamma z^{-\gamma} \tag{5.14}$$

for appropriate choices of $c_\gamma \in \mathcal{C}[\mathcal{A}]$. Note that $c_0 = \mathcal{P}_B(c)$. Hence it is sufficient to show that each c_γ can be expressed as a linear combination of the $\tau(\beta_i)$ for $i = 1, \ldots, n$. We proceed by induction on restricted height. Assume that c_γ can be expressed as a linear combination of the $\tau(\beta_i)$ for all γ of restricted height strictly less than m. Let η be a restricted weight such that $\mathrm{ht}_r(\eta) = m$.

Suppose that $\lambda \in P^+(\Sigma)$. By (5.12), there exist scalars $r_{\lambda\gamma}$ with $r_{\lambda 0} = 1$ such that

$$\varphi_{2\lambda} = \sum_{\beta \geq 0} r_{\lambda\beta} z^{2\lambda - \beta}. \tag{5.15}$$

Expressions (5.14) and (5.15) combined with Theorem 5.4 yield

$$z^{2\lambda}(\mathcal{P}_B(c))\varphi_{2\lambda} = \Big(\sum_{\beta \leq \lambda} r_{\lambda\beta} z^{2\lambda-\beta}\Big) * \Big(\sum_{\gamma \leq 0} c_\gamma z^{-\gamma}\Big). \tag{5.16}$$

Analyzing the coefficient of $z^{2\lambda - \eta}$ in (5.16) yields

$$z^{2\lambda}(\mathcal{P}_B(c)) r_{\lambda\eta} = z^{2\lambda}(c_\eta) + \sum_{\eta \geq \beta > 0} r_{\lambda\beta} z^{2\lambda-\beta}(c_{\eta-\beta}). \tag{5.17}$$

It follows from the defining relation (5.1) of $\mathcal{C}[Q(\Sigma)]\mathcal{A}$ that $z^{-\gamma}\tau(\beta)z^{\gamma} = q^{-(\gamma,\beta)}\tau(\beta)$ for all $\tau(\beta) \in \mathcal{A}$ and $\gamma \in Q(\Sigma)$. Moreover, (1.18) ensures that $z^{2\lambda-\beta}(a) = z^{2\lambda}(z^{-\beta} a z^{\beta})$ for all $a \in \mathcal{C}[\mathcal{A}]$. Now the inductive hypothesis implies that $z^{-\beta}c_{\eta-\beta}z^{\beta}$ is a linear combination of $\tau(\beta_i)$, for $i = 1, \ldots, n$ and $\beta > 0$. The lemma now follows from the fact that we can use (5.17) to express c_η as a linear combination of $\mathcal{P}_B(c)$ and the $c_{\eta-\beta}$ for $\eta \geq \beta > 0$. More precisely, (5.17) implies that

$$z^{2\lambda}(c_\eta) = z^{2\lambda}\Big(r_{\lambda\eta}\mathcal{P}_B(c) - \sum_{\eta \geq \beta > 0} r_{\lambda\beta} z^{-\beta} c_{\eta-\beta} z^{\beta}\Big)$$

for all $\lambda \in P^+(\Sigma)$. Hence

$$c_\eta = r_{\lambda\eta}\mathcal{P}_B(c) - \sum_{\eta \geq \beta > 0} r_{\lambda\beta} z^{-\beta} c_{\eta-\beta} z^{\beta}.$$

□

We can define a degree function on $\mathcal{C}[\mathcal{A}]$ using the restricted height function (see (5.13)). In particular, given an element $X = \sum_\beta a_\beta \tau(\beta)$ of $\mathcal{C}[\mathcal{A}]$, set

$$\deg(X) = \max\{\mathrm{ht}_r(\beta) |\ a_\beta \neq 0\}. \tag{5.18}$$

We can further extend this degree function to a degree function on $\mathcal{C}(Q(\Sigma))\mathcal{A}$ by insisting that nonzero elements of $\mathcal{C}(Q(\Sigma))$ have degree 0. Given $X \in \mathcal{C}(Q(\Sigma))\mathcal{A}$, let $\mathrm{top}(X)$ denote the element of $\mathcal{C}[\mathcal{A}]$ homogeneous of degree $\deg(X)$ such that $X - \mathrm{top}(X)$ has degree strictly less than $\deg(X)$.

The next result uses the close connection between $\mathcal{X}$ and $\mathcal{P}_B$ to gain further information about the image of $\check{U}^B$ under $\mathcal{P}_B$.

THEOREM 5.9. *The set* $\{\text{top}(X)|X \in \mathcal{P}_B(\check{U}^B)\}$ *is a subset of the* $\mathcal{C}$ *span of* $\{\tau(2\eta)|\eta \in P^+(\Sigma)\}$.

PROOF. Let $c \in \check{U}^B$ and write $\mathcal{P}_B(c) = \sum_{i=1}^n a_{\beta_i}\tau(\beta_i)$ where each a_{β_i} is a nonzero scalar. By Lemma 5.8, we can write

$$\mathcal{X}(c) = \sum_{i=1}^{n} f_{\beta_i}\tau(\beta_i)$$

where $f_{\beta_i} \in \mathcal{C}[[Q^-(\Sigma)]]$. Lemma 5.8 further ensures that the constant term of each f_{β_i} is nonzero. By Theorem 5.5, $\mathcal{X}(c)$ is W_Θ invariant. Hence, the set $\{\tau(\beta_i)|\ 1 \leq i \leq n\}$ is W_Θ invariant. Therefore $\text{top}(\mathcal{P}_B(c))$ is a linear combination of those $\tau(\beta_i)$ such that β_i a dominant integral restricted weight. $\square$

CHAPTER 6

The Image of the Center

In this section, we study the center of $\check{U}$ under the map $\mathcal{P}_B$, for $B \in \mathcal{B}$. Using the ordinary Harish-Chandra map $\mathcal{P}$ defined in Section 3, we show that $\mathcal{P}_B(Z(\check{U}))$ is invariant under a dotted action of the restricted Weyl group W_Θ. We further obtain information about possible highest degree terms of certain elements in $\mathcal{P}_B(Z(\check{U}))$. This information combined with Theorem 5.9 and Theorem 2.6 yields that $\mathcal{P}_B(Z(\check{U}))$, and hence $\mathcal{P}_B(\check{U}^B)$, is equal to the entire invariant ring of $\mathcal{C}[\mathcal{A}]$ under the dotted action of W_Θ whenever $\widetilde{P^+(\pi)} = P^+(\Sigma)$. We further show that this result fails if $\mathfrak{g}, \mathfrak{g}^\theta$ is of type EIII, EIV, EVII, or EIX.

Recall the definition of the two different types of Harish-Chandra maps $\mathcal{P}$ (beginning of Section 3) and $\mathcal{P}_B$ (Definition 4.3). Recall further the projection $\tilde{\mathcal{P}}$ of $\check{U}^0$ onto $\mathcal{C}[\check{\mathcal{A}}]$ defined using the direct sum decomposition in (3.5). It follows directly from the definition of these maps that $\mathcal{P}_B \circ \mathcal{P}$ is equal to $\tilde{\mathcal{P}} \circ \mathcal{P}$ for all $B \in \mathcal{B}$. In general, $\mathcal{P}_B \circ \mathcal{P}$ does not agree with $\mathcal{P}_B$. However, the next lemma shows that these two maps agree upon restriction to the center $Z(\check{U})$. In particular, the restriction of $\mathcal{P}_B$ to $Z(\check{U})$ is *independent* of the choice of $B \in \mathcal{B}$.

LEMMA 6.1. *The restriction of $\mathcal{P}_B \circ \mathcal{P}$ to $Z(\check{U})$ agrees with the restriction of $\mathcal{P}_B$ to $Z(\check{U})$.*

PROOF. Suppose that $\lambda \in P^+(\Sigma)$. Recall that the zonal spherical function $g_{2\lambda}$ is an element of the U bimodule $L(2\lambda) \otimes L(2\lambda)^*$ (see Section 1). Now elements of $Z(\check{U})$ act as scalars (on both the right and left) on $L(2\lambda) \otimes L(2\lambda)^*$ via the central character corresponding to the dominant integral weight 2λ. In particular, we have

$$g_{2\lambda} \cdot c = z^{2\lambda}(\mathcal{P}(c))g_{2\lambda}$$

for all $c \in Z(\check{U})$ and $\lambda \in P^+(\Sigma)$. On the other hand, given $\beta \in P(\Sigma)$, we have that

$$z^\beta(\mathcal{P}(u)) = z^\beta(\mathcal{P}_B \circ \mathcal{P}(u))$$

for all $u \in \check{U}$. Therefore,

$$g_{2\lambda} \cdot c = z^{2\lambda}(\mathcal{P}_B \circ \mathcal{P}(c))g_{2\lambda}$$

for all $c \in Z(\check{U})$ and $\lambda \in P^+(\Sigma)$. By (5.6) we further have

$$z^{2\lambda}(\mathcal{P}_B(c))g_{2\lambda} = z^{2\lambda}(\mathcal{P}_B \circ \mathcal{P}(c))g_{2\lambda}$$

for all $c \in Z(\check{U})$ and $\lambda \in P^+(\Sigma)$. The lemma now follows from the facts that both $\mathcal{P}_B(c)$ and $\mathcal{P}_B \circ \mathcal{P}(c)$ are elements of $\mathcal{C}[\mathcal{A}]$ and $P^+(\Sigma)$ is a Zariski dense subset of $\mathfrak{a}^*$. □

The next result is an immediate consequence of the previous lemma, Theorem 3.1, and Theorem 3.3.

THEOREM 6.2. *For each* $f \in Z(\check{U})$, $\mathcal{P}_B(f)$ *is invariant under the dotted action of* W_Θ. *Moreover,* $\mathcal{P}_B(Z(\check{U}))$ *is a subring of* $\mathcal{C}[\mathcal{A}]^{W_\Theta \circ}$.

A description of a nice basis for the center of $Z(\check{U})$ and the image of this basis under $\mathcal{P}$ is given in [Jo, Chapter 7] and [JL1]. It should be noted that these references use the locally finite part of $\check{U}$ with respect to the left adjoint action. Since we are viewing the center as a subalgebra of the $(\text{ad}_r\ B)$ invariants of $\check{U}$, it is necessary to translate these results to the setting of the right adjoint action. In particular, for each $\mu \in P^+(\pi)$ there exists a unique central element $z_{2\mu}$ in $\tau(2\mu) + (\text{ad}_r\ U_+)\tau(2\mu)$. Moreover, the set $\{z_{2\mu} | \mu \in P^+(\pi)\}$ forms a basis for $Z(\check{U})$. Given $\lambda \in P^+(\pi)$, set

$$\hat{\tau}(\lambda) = \sum_{w \in W} \tau(w\lambda) q^{(\rho, w\lambda)}. \tag{6.1}$$

The next lemma, which describes the image of each of these basis elements under $\mathcal{P}$, is a version of [Jo, Lemma 7.1.19] with respect to the right adjoint action.

LEMMA 6.3. *For all* $\mu \in P^+(\pi)$,

$$\mathcal{P}(z_{2\mu}) = a_{2\mu}^{-1} \sum_{\nu \in P^+(\pi)} \hat{\tau}(2\nu) \dim L(\mu)_\nu$$

where

$$a_{2\mu} = \sum_{\nu \in P^+(\pi)} \Big(\sum_{w \in W} q^{(\rho, 2w\nu)} \Big) \dim L(\mu)_\nu.$$

PROOF. Let ι be the $\mathbf{C}$ algebra involutive antiautomorphism of $\check{U}$ defined by $\iota(x_i) = y_i$ and $\iota(y_i) = x_i$ for all $1 \leq i \leq n$, $\iota(q) = q^{-1}$, and $\iota(t) = t^{-1}$ for all $t \in \check{U}$. Using the formulas in (1.14) and (1.15) for the left and right adjoint actions, it is straightforward to check that

$$\iota((\text{ad}\ a)b) = -(\text{ad}_r\ \iota(a))\iota(b) \tag{6.2}$$

for all $a \in \{x_i, y_i |\ 1 \leq i \leq n\} \cup \check{T}$ and $b \in \check{U}$. Hence (6.2) holds for all $a \in \check{U}$ and $b \in \check{U}$. It follows that

$$\iota((\text{ad}\ U_+)\tau(-2\mu)) = (\text{ad}_r\ U_+)\tau(2\mu)$$

for all $\mu \in P^+(\pi)$. Thus $\iota(z_{2\mu})$ is the unique central element contained in $\tau(-2\mu) + (\text{ad}\ U_+)\tau(-2\mu)$. Since ι preserves both $\check{U}^0$ and $G_+^-\check{U} + \check{U}U_+^+$, it follows that ι commutes with the map $\mathcal{P}$. In particular $\mathcal{P}(z_{2\mu}) = \iota(\mathcal{P}(\iota(z_{2\mu})))$. Thus by [Jo, Lemma 7.1.19], we have

$$\mathcal{P}(z_{2\mu}) = a_{2\mu}^{-1} \sum_{\nu \in P^+(\pi)} \hat{\tau}(2\nu) \dim L(\mu)_\nu$$

where $a_{2\mu}$ is a nonzero scalar.

Recall that ϵ is the counit of the Hopf algebra $\check{U}$. Now $\tau(2\mu) \in 1 + \check{U}_+$ and $(\text{ad}_r\ U_+)\tau(2\mu) \in \check{U}_+$. It follows that $\epsilon(z_{2\mu}) = \epsilon(1) = 1$. On the other hand, $z^0(\mathcal{P}(z_{2\mu})) = \epsilon(\mathcal{P}(z_{2\mu})) = \epsilon(z_{2\mu})$. Hence $z^0(\mathcal{P}(z_{2\mu})) = 1$ and so

$$a_{2\mu} = z^0\Big(\sum_{\nu \in P^+(\pi)} \hat{\tau}(2\nu) \dim L(\mu)_\nu \Big).$$

□

Now suppose that $\mu \in P^+(\pi)$. Temporarily set $a = a_{2\mu}$ where $a_{2\mu}$ is the scalar defined in the previous lemma for the central element $z_{2\mu}$. Using Lemma 6.3, we can write $a\mathcal{P}(z_{2\mu})$ as a sum

$$b_\mu \tau(2\mu) + \sum_{\gamma<\mu} b_\gamma \tau(2\gamma)$$

where the b_γ are scalars. Since $\dim L(\mu)_\mu = 1$, it follows that b_μ is just a power of q times the order of the stabilizer of μ in W. More generally, $b_\gamma \in \mathbf{N}[q, q^{-1}]$ for each γ. By Lemma 6.1, $\mathcal{P}_B(z_{2\mu}) = \mathcal{P}_B \circ \mathcal{P}(z_{2\mu})$. Hence

$$a\mathcal{P}_B(z_{2\mu}) \in (\sum_{\tilde{\gamma}=\tilde{\mu}} b_\gamma)\tau(2\tilde{\mu}) + \sum_{\tilde{\gamma}<\tilde{\mu}} \mathcal{C}\tau(2\tilde{\gamma}). \tag{6.3}$$

It follows that the sum $\sum_{\tilde{\gamma}=\tilde{\mu}} b_\gamma$ is nonzero. Recall the notion of top degree homogeneous term, denoted by $\text{top}(X)$ for $X \in \mathcal{C}[\mathcal{A}]$, given in Section 5 (see (5.18) and subsequent discussion). The above discussion ensures that $\text{top}(\mathcal{P}_B(z_{2\mu}))$ is a nonzero multiple of $\tau(2\tilde{\mu})$. Hence

$$\text{span}\{\tau(2\tilde{\mu})|\ \mu \in P^+(\pi)\} \subseteq \{\text{top}(\mathcal{P}_B(z))|\ z \in Z(\check{U})\}. \tag{6.4}$$

The next lemma provides a sufficient, but not necessary condition for $\mathcal{P}_B(Z(\check{U}))$ to equal $\mathcal{P}_B(\check{U}^B)$.

LEMMA 6.4. *If* $\widetilde{P^+(\pi)} = P^+(\Sigma)$ *then* $\mathcal{P}_B(Z(\check{U})) = \mathcal{P}_B(\check{U}^B)$.

PROOF. By Theorem 5.9, the set $\{\text{top}(\mathcal{P}_B(u))|\ u \in \check{U}^B\}$ is a subset of the span of the set $\{\tau(2\mu)|\ \mu \in P^+(\Sigma)\}$. The lemma now follows from (6.4). □

Given $\mu \in P^+(\Sigma)$, set

$$\hat{m}(2\mu) = \sum_{\gamma \in W_\Theta \mu} q^{(\tilde{\rho}, 2\gamma)}\tau(2\gamma). \tag{6.5}$$

Note that $\hat{m}(2\mu)$ is an element of $\mathcal{C}[\mathcal{A}]^{W_{\Theta^\circ}}$. Moreover, the set $\{\hat{m}(2\mu)|\mu \in P^+(\Sigma)\}$ forms a basis for $\mathcal{C}[\mathcal{A}]^{W_{\Theta^\circ}}$. Now $\text{top}(\hat{m}(2\mu)) = q^{(\tilde{\rho}, 2\mu)}\tau(2\mu)$. Hence

$$\{\text{top}(u)|\ u \in \mathcal{C}[\mathcal{A}]^{W_{\Theta^\circ}}\} = \text{span}\{\tau(2\mu)|\ \mu \in P^+(\Sigma)\}. \tag{6.6}$$

The next result establishes Theorem B of the introduction for those symmetric pairs that do not contain a symmetric pair of type AII.

THEOREM 6.5. *Suppose that* $\mathfrak{g}, \mathfrak{g}^\theta$ *is not of type EIII, EIV, EVII, EIX or CII(2) and* $\mathfrak{g}$ *does not contain a* θ *invariant Lie subalgebra* $\mathfrak{r}$ *of rank greater than or equal to 7 such that* $\mathfrak{r}, \mathfrak{r}^\theta$ *is of type AII. Then*

$$\mathcal{P}_B(Z(\check{U})) = \mathcal{P}_B(\check{U}^B) = \mathcal{C}[\mathcal{A}]^{W_{\Theta^\circ}}.$$

PROOF. Assume $\mathfrak{g}, \mathfrak{g}^\theta$ is not of type FII. By Theorem 2.6, $\widetilde{P^+(\pi)} = P^+(\Sigma)$. Hence Theorem 5.9 and (6.4) imply that

$$\begin{aligned}\text{span}\{\tau(2\mu)|\ \mu \in P^+(\Sigma)\} =& \{\text{top}(u)|\ u \in \mathcal{P}_B(Z(\check{U}))\}\\ =& \{\text{top}(u)|\ u \in \mathcal{P}_B(\check{U}^B)\}.\end{aligned}$$

On the other hand, by Theorem 6.2, $\mathcal{P}_B(Z(\check{U}))$ is a subring of $\mathcal{C}[\mathcal{A}]^{W_{\Theta^\circ}}$. The theorem now follows in this case from (6.6).

Now assume $\mathfrak{g}, \mathfrak{g}^\theta$ is of type FII. Recall the notation of Section 2 and Lemma 2.5. The character formulas for the finite-dimensional simple modules $L(\omega_1)$ and $L(\omega_4)$ combined with Lemma 6.3 yields

$$\mathcal{P}(z_{2\omega_1})) = a_1[\hat{\tau}(2\omega_1) + \hat{\tau}(2\omega_4) + 4]$$

and

$$\mathcal{P}(z_{2\omega_4}) = a_4(\hat{\tau}(2\omega_4) + 2)$$

for nonzero scalars a_1 and a_4. By Lemma 2.5, $\tilde{\omega}_1 = 2\omega_4' = \tilde{\omega}_4$. A straightforward computation yields

$$\mathcal{P}_B(\hat{\tau}(2\omega_1)) = \mathcal{P}_B \circ \mathcal{P}(\hat{\tau}(2\omega_1)) = 6\hat{m}(4\omega_4') + 12$$

and

$$\mathcal{P}_B(\hat{\tau}(2\omega_4)) = \mathcal{P}_B \circ \mathcal{P}(\hat{\tau}(2\omega_4)) = \hat{m}(4\omega_4') + 8\hat{m}(2\omega_4') + 6.$$

It follows that $\hat{m}(2\omega_4')$ can be written as a linear combination of $\mathcal{P}_B(z_{2\omega_1})$, $\mathcal{P}_B(z_{2\omega_4})$, and 1. This case now follows from Theorem 5.9, Theorem 6.2 and the fact that $\mathcal{C}[\mathcal{A}]^{W_{\Theta^\circ}}$ is generated by $\hat{m}(2\omega_4')$. □

The proof of Theorem 6.5 uses the fact that $\widetilde{P^+(\pi)} = P^+(\Sigma)$ for most of the symmetric pairs under consideration. We see in Section 8 that the conclusion of Theorem 6.5 holds for another family of symmetric pairs. On the other hand, we show below that for symmetric pairs $\mathfrak{g}, \mathfrak{g}^\theta$ of type EIII, EIV, EVII, and EIX, the set $\mathcal{P}_B(Z(\check{U}))$ is a proper subset of $\mathcal{C}[\mathcal{A}]^{W_\Theta}$. In particular, the above theorem cannot be extended to these symmetric pairs.

Let $\check{U}_{\mathbf{C}(q)}$ denote the $\mathbf{C}(q)$ subalgebra of $\check{U}$ generated by x_i, y_i for $1 \leq i \leq n$ and the group $\check{T}$. Now the center $Z(\check{U})$ is isomorphic to a polynomial ring in n variables. Moreover, we may choose the generators of $Z(\check{U})$ to be elements of the smaller algebra $\check{U}_{\mathbf{C}(q)}$ as in [Jo, Section 7]. In particular, we may restrict our attention to the $\mathbf{C}(q)$ algebra $\check{U}_{\mathbf{C}(q)}$.

The next result is proved using specialization techniques and classical results. First we recall basic notions concerning specialization. Set A equal to the localization $\mathbf{C}[q]_{(q-1)}$ of $\mathbf{C}[q]$ at the maximal ideal $(q-1)$. Let $\hat{U}$ denote the A subalgebra of $\check{U}_{\mathbf{C}(q)}$ generated by x_i, y_i for $1 \leq i \leq n$, $\check{T}$, and $(t-1)/(q-1)$ for all $t \in \check{T}$. There is an algebra isomorphism

$$\hat{U} \otimes_A \mathbf{C} \to U(\mathfrak{g}) \tag{6.7}$$

which sends each $x_i \otimes 1$ to the root vector e_i of weight α_i and each $y_i \otimes 1$ to the root vector f_i of weight $-\alpha_i$ (see for example [L1, Section 2]). Given a subalgebra S of $\check{U}$, we say that S specializes to the subalgebra $\bar{S}$ of $U(\mathfrak{g})$ provided that the image of $S \cap \hat{U}$ in $U(\mathfrak{g})$ is $\bar{S}$.

To make the exposition easier, we assume that $\mathfrak{g}$ is simply laced and so all roots have the same length (which we assume is 2). Ultimately, our focus will be on the four exceptional symmetric pairs where the underlying Lie algebra $\mathfrak{g}$ is simply laced of type E_6, E_7, or E_8. So this additional assumption does not affect the next theorem.

Recall (Section 1) that $h_1, \ldots, h_n$ is a basis of coroots for the Cartan subalgebra $\mathfrak{h}$ of $\mathfrak{g}$. We have that $(t_i - 1)/(q-1)$ is sent to h_i for all $1 \leq i \leq n$ under the isomorphism (6.7). More generally, if $\beta = \sum_i m_i\alpha_i$, then $(\tau(\beta) - 1)/(q-1)$ specializes to $\sum_i m_i h_i$. Note further that t specializes to 1 for each $t \in \check{T}$. Set

$T_2 = \{\tau(2\mu) | \ \mu \in P(\pi)\}$. Note that $(\tau(2\alpha_i) - 1)(q-1)^{-1}$ specializes to $2h_i$. It follows that the specialization of $\mathcal{C}[T_2]$ at $q = 1$ is just the enveloping algebra $U(\mathfrak{h})$ of the Cartan subalgebra $\mathfrak{h}$.

Let $\mathfrak{a}$ denote the eigenspace for θ with eigenvalue -1 inside $\mathfrak{h}$. It is straightforward to see that the subspace $\mathfrak{a}$ of $\mathfrak{h}$ equals the set $\{h - \theta(h) | h \in \mathfrak{h}\}$. Note that the subspace $\mathfrak{a}^*$ defined in Section 3 can be identified with the dual of $\mathfrak{a}$. We can define a projection map $\tilde{}$ from $\mathfrak{h}$ onto $\mathfrak{a}$ analagous to the projection map $\tilde{}$ from $\mathfrak{h}^*$ onto $\mathfrak{a}^*$ as follows. Set $\tilde{h} = (h - \theta(h))/2$ for all $h \in \mathfrak{h}$. Note that this projection extends in the obvious way to an algebra homomorphism of $U(\mathfrak{h})$ onto $U(\mathfrak{a})$. Now the image of $\mathcal{C}[\mathcal{A}]$ under specialization is just $U(\mathfrak{a})$. Recall that the restriction of $\mathcal{P}_B$ to $\check{U}^0$ sends $\tau(\mu)$ to $\tau(\tilde{\mu})$ for all $\mu \in P(\pi)$. Also, Lemma 2.2 ensures that the image of $\mathcal{C}[T_2]$ under $\mathcal{P}_B$ is a subalgebra of $\mathcal{C}[\mathcal{A}]$. In particular, we have a commutative diagram:

$$\begin{array}{ccc} \mathcal{C}[T_2] & \rightarrow & U(\mathfrak{h}) \\ \downarrow & & \downarrow \\ \mathcal{C}[\mathcal{A}] & \rightarrow & U(\mathfrak{a}) \end{array} \tag{6.8}$$

where the top and bottom maps are specialization and the downward maps are algebra homomorphisms defined using $\tilde{}$. Moreover, all maps in this diagram are surjections.

Let α be a simple root in π and let s_α denote the corresponding reflection in W. For each i such that $1 \leq i \leq n$, we have

$$\begin{aligned} s_\alpha \cdot \frac{(t_i - 1)}{(q-1)} &= \frac{(q^{(\rho, s_\alpha \alpha_i - \alpha_i)}\tau(s_\alpha \alpha_i) - 1)}{(q-1)} \\ &= \frac{(q^{(\rho, s_\alpha \alpha_i - \alpha_i)} - 1)}{(q-1)}\tau(s_\alpha \alpha_i) + \frac{(\tau(s_\alpha \alpha_i) - 1)}{(q-1)}. \end{aligned}$$

It follows that $s_\alpha \cdot (t_i - 1)(q-1)^{-1}$ specializes to $s_\alpha h_i + (\rho, s_\alpha \alpha_i) - (\rho, \alpha_i)$. In particular, the dotted action of W on $\check{U}^0$ specializes to the dotted (or translated) action of W on $U(\mathfrak{h})$ (see [H, Section 23.3]). A similar argument shows that the dotted action of W_Θ on $\mathcal{C}[\mathcal{A}]$ specializes to the dotted action of W_Θ on $U(\mathfrak{a})$ as defined in [He, Section 2]. Thus we have a commutative diagram

$$\begin{array}{ccc} \mathcal{C}[T_2]^{W\cdot} & \rightarrow & U(\mathfrak{h})^{W\cdot} \\ \downarrow & & \downarrow \\ \mathcal{C}[\mathcal{A}]^{W_\Theta \circ} & \rightarrow & U(\mathfrak{a})^{W_\Theta \circ} \end{array} \tag{6.9}$$

where the maps are the same as in the commutative diagram (6.8). Note that the top and bottom maps are still surjections.

THEOREM 6.6. *Suppose $\mathfrak{g}, \mathfrak{g}^\theta$ is an irreducible symmetric pair of type EIII, EIV, EVII, or EIX. Then $\mathcal{P}_B(Z(\check{U})) \neq \mathcal{C}[\mathcal{A}]^{W_\Theta \circ}$*

PROOF. Note that the image of the center $Z(\check{U})$ under the ordinary (quantum) Harish-Chandra map $\mathcal{P}$ is just $\mathcal{C}[T_2]^{W\cdot}$ (Theorem 3.1). Assume that $\mathcal{P}_B(Z(\check{U})) = \mathcal{C}[\mathcal{A}]^{W_\Theta \circ}$. Recall that the restriction of $\mathcal{P}_B$ to $Z(\check{U})$ agrees with $\mathcal{P}_B \circ \mathcal{P}$. It follows that the map from $\mathcal{C}[T_2]^{W\cdot}$ to $\mathcal{C}[\mathcal{A}]^{W_\Theta \circ}$ in the commutative diagram (6.9) is surjective. Hence the map from $U(\mathfrak{h})^{W\cdot}$ to $U(\mathfrak{a})^{W_\Theta \circ}$ is surjective. Therefore the image of the center of $U(\mathfrak{g})$ under the classical Harish-Chandra map associated to $\mathfrak{g}, \mathfrak{g}^\theta$ is equal to the entire invariant ring $U(\mathfrak{a})^{W_\Theta \circ}$. This contradicts [He]. $\square$

It should be noted that surjectivity of the map from $U(\mathfrak{h})^{W\cdot}$ to $U(\mathfrak{a})^{W_\Theta \circ}$ does not imply surjectivity of the map from $\mathcal{C}[T_2]^{W\cdot}$ to $\mathcal{C}[\mathcal{A}]^{W_\Theta \circ}$. The problem is that there are many subrings of $\mathcal{C}[\mathcal{A}]^{W_\Theta \circ}$ which surject onto $U(\mathfrak{a})^{W_\Theta \circ}$. (One such example is $\mathcal{C}[t^2 | t \in \mathcal{A}]^{W_\Theta \circ}$.) The reader is referred to [JL2, 6.13] for a similar situation. In particular, it is shown in [JL2] that the center $Z(U)$ of the ordinary quantized enveloping algebra U specializes to the center $Z(\mathfrak{g})$ of $U(\mathfrak{g})$. However, for many simple Lie algebras $\mathfrak{g}$, $Z(U)$ is not a polynomial ring while $Z(\mathfrak{g})$ is always a polynomial ring.

CHAPTER 7

Finding Invariant Elements

Recall that the ordinary Harish-Chandra map is a projection of $\check{U}$ onto $\check{U}^0$. Furthermore, this Harish-Chandra map induces an isomorphism between $Z(\check{U})$ and a particular subring of $\check{U}^0$. There are a number of different proofs of this result in the literature (see [B]). The approach taken by [JL1] (see also [Jo, Section 7]) establishes the surjectivity of this isomorphism by using a map which projects elements of the intersection $\check{T} \cap F(\check{U})$ onto $Z(\check{U})$. In this section, we study an analog $\mathcal{L}$ of this map associated to quantum symmetric pairs. Some of the material presented here is a generalization of parts of [L5, Section 7].

Fix a subalgebra B in $\mathcal{B}$. Let ϕ be the Hopf algebra automorphism of U which fixes elements in T and satisfies

$$\phi(x_i) = q^{(-2\rho,\tilde{\alpha}_i)}x_i \qquad \phi(y_i) = q^{(2\rho,\tilde{\alpha}_i)}y_i \tag{7.1}$$

for all $1 \leq i \leq n$. Note that $\tilde{\alpha}_i = 0$ whenever $\Theta(\alpha_i) = \alpha_i$. Thus ϕ acts as the identity on the subalgebra $\mathcal{M}$ of U. Now B is generated by $\mathcal{M}, T_\Theta$ and the B_i for $\alpha_i \notin \pi_\Theta$ (Section 1). Furthermore, $B_i = y_it_i + d_i\tilde{\theta}(y_i)t_i + s_it_i$ for suitably chosen scalars d_i and s_i. Set $C_i = B_i - s_i$ for each $\alpha_i \notin \pi_\Theta$. Note that $\epsilon(C_i) = 0$ where ϵ is the counit of $\check{U}$. Clearly, B is generated by $\mathcal{M}, T_\Theta$, and the C_i for $\alpha_i \notin \pi_\Theta$. Now for most α_i, in particular, when $\alpha_i \notin \mathcal{S}$, we have that C_i simply equals B_i. More generally, the advantage of replacing B_i by C_i when $s_i \neq 0$ is that C_i is in B_+.

Recall the definitions of the antipode σ and the comultiplication map Δ given in Section 1 ((1.11), (1.12), and (1.13)). The following lemma generalizes [L5, Lemma 7.3].

LEMMA 7.1. *Suppose that $a \in \check{U}$ and $b \in B_+$. Then $(\mathrm{ad}_r\ b)a$ is contained in $\phi(B_+)\check{U} + \check{U}B_+$.*

PROOF. Note that $\mathcal{M}T_\Theta$ is a Hopf subalgebra of U and a subalgebra of both $\phi(B)\check{T}_\Theta$ and $B\check{T}_\Theta$. Hence

$$\begin{aligned}(\mathrm{ad}_r\ (\mathcal{M}T_\Theta)_+)a &\subseteq (\mathcal{M}T_\Theta)_+a\mathcal{M}T_\Theta + \mathcal{M}T_\Theta a(\mathcal{M}T_\Theta)_+ \\ &\subseteq (\phi(B_+))a + \check{U}B_+.\end{aligned}$$

It remains to check that $(\mathrm{ad}_r\ C_i)a$ is contained in $\phi(B_+)\check{U} + \check{U}B_+$ for each i such that $\alpha_i \in \pi \setminus \pi_\Theta$. To do this, we examine the image of y_it_i, $s_i(t_i - 1)$, and $\tilde{\theta}(y_i)$ under both the coproduct Δ and the antipode σ. Note that

$$\Delta(y_it_i) = y_it_i \otimes 1 + t_i \otimes y_it_i$$

and

$$\Delta(t_i - 1) = t_i \otimes t_i - 1 \otimes 1 = (t_i - 1) \otimes 1 + t_i \otimes (t_i - 1)$$

for each i. By the proof of [L5, Lemma 7.3] we have that

$$\Delta(\tilde{\theta}(y_i)t_i) \in \tilde{\theta}(y_i)t_i \otimes \tau(\Theta(\alpha_i))t_i + t_i \otimes \tilde{\theta}(y_i)t_i + U \otimes (\mathcal{M}T_\Theta)_+.$$

Note further that $\tau(\Theta(\alpha_i))t_i - 1 \in \mathcal{C}[T_\Theta]_+$. It follows that

$$\Delta(C_i) \in C_i \otimes 1 + t_i \otimes C_i + U \otimes (\mathcal{M}T_\Theta)_+. \tag{7.2}$$

We turn our attention to the antipode. It follows easily from the definition of σ that

$$\sigma(y_i t_i) = -t_i^{-1} y_i t_i = -q^{(2\rho,\alpha_i)} y_i$$

and

$$\sigma(t_i - 1) = (t_i^{-1} - 1)$$

for each i. By the proof of [L5, proof of Lemma 7.3], we have that

$$\sigma(\tilde{\theta}(y_i)t_i) \in (\mathcal{M}T_\Theta)_+ U - q^{(2\rho,\Theta(\alpha_i))}\tilde{\theta}(y_i).$$

Hence

$$\begin{aligned}\sigma(C_i) &\in -q^{(2\rho,\alpha_i)} y_i - d_i q^{(2\rho,\Theta(\alpha_i))}\tilde{\theta}(y_i) + s_i(t_i^{-1} - 1) + (\mathcal{M}T_\Theta)_+ U \\ &= q^{(\rho,\Theta(\alpha_i)+\alpha_i)}\phi(C_i)t_i^{-1} + (\mathcal{M}T_\Theta)_+ U.\end{aligned} \tag{7.3}$$

By (7.2) and (7.3) we see that

$$\begin{aligned}(\mathrm{ad}_r\ C_i)a &\in -\sigma(C_i)a + t_i^{-1}aC_i + Ua(\mathcal{M}T_\Theta)_+ \\ &\subseteq -q^{(\rho,\Theta(\alpha_i)+\alpha_i)}\phi(C_i)t_i^{-1}a + (\mathcal{M}T_\Theta)_+ Ua + t_i^{-1}aC_i + \check{U}(\mathcal{M}T_\Theta)_+ \\ &\subseteq \phi(B_+)\check{U} + \check{U}B_+\end{aligned}$$

for all $a \in \check{U}$. □

Recall the definition of the Hopf algebra automorphism χ of U given in (1.27). Note that $\phi = \chi^{-2}$. Note further that $(\rho, \tilde{\alpha}_i) = (\tilde{\rho}, \tilde{\alpha}_i) = (\tilde{\rho}, \alpha_i)$. Hence $\tau(\tilde{\rho})x_i\tau(-\tilde{\rho}) = \chi(x_i)$ and $\tau(\tilde{\rho})y_i\tau(-\tilde{\rho}) = \chi(y_i)$ for all $1 \le i \le n$. It follows that $\tau(\tilde{\rho})u\tau(-\tilde{\rho}) = \chi(u)$ for all $u \in U$.

LEMMA 7.2. *We have the following equality of sets:*

$$\tau(\tilde{\rho})(\phi(B_+)\check{U} + \check{U}B_+) = \chi^{-1}(B_+)\tau(\tilde{\rho})\check{U} + \tau(\tilde{\rho})\check{U}B_+.$$

PROOF. By (7.1), we have that $\phi(u) = \tau(-2\tilde{\rho})u\tau(2\tilde{\rho})$ for all $u \in U$. Hence

$$\tau(\tilde{\rho})\phi(B)\tau(-\tilde{\rho}) = \tau(-\tilde{\rho})B\tau(\tilde{\rho}) = \chi^{-1}(B).$$

It follows that $\tau(\tilde{\rho})(\phi(B))\check{U} = \chi^{-1}(B)\tau(\tilde{\rho})\check{U}$. □

As explained in Section 1, $\check{U}^B$ is a subspace of the locally finite part $F_r(\check{U})$ of $\check{U}$. Furthermore, the action of $(\mathrm{ad}_r\ B)$ on $F_r(\check{U})$ is locally finite and semisimple (see Theorem 1.2). It follows from the discussion concerning (1.23) that

$$(\mathrm{ad}_r\ B_+)F_r(\check{U}) \cap \check{U}^B = 0.$$

Moreover,

$$F_r(\check{U}) = \check{U}^B \oplus (\mathrm{ad}_r\ B_+)F_r(\check{U}). \tag{7.4}$$

Now suppose that $a \in F_r(\check{U})$ but $a \notin (\mathrm{ad}_r\ B_+)\check{U}$. We have that $(\mathrm{ad}_r\ B)a$ is a finite-dimensional submodule of $F_r(\check{U})$. By Theorem 1.2, $(\mathrm{ad}_r\ B)a$ can be written as a direct sum of simple $(\mathrm{ad}_r\ B)$ modules. Moreover, $(\mathrm{ad}_r\ B_+)a$ is a finite-dimensional $(\mathrm{ad}_r\ B)$ submodule of $(\mathrm{ad}_r\ B)a$ with codimension at most 1. In particular, there exists $a' \in \check{U}^B$ such that

$$(\mathrm{ad}_r\ B)a = \mathcal{C}a' \oplus (\mathrm{ad}_r\ B_+)a.$$

The next definition is based on [L5, Section 7, before Lemma 7.5].

DEFINITION 7.3. The map $\mathbf{L}$ is the projection map from $F_r(\check{U})$ onto $\check{U}^B$ using the direct sum decomposition (7.4). Moreover, given $a \in F_r(\check{U})$ such that $a \notin (\mathrm{ad}_r\ B_+)F_r(\check{U})$, we have that $\mathbf{L}(a)$ is the unique element of $\check{U}^B$ which is contained in $a + (\mathrm{ad}_r\ B_+)a$.

Let κ denote the conjugate linear antiautomorphism of $\check{U}$ defined in Section 1. As discussed in Section 1, κ is chosen so that $\kappa(B) = B$. Moreover $\kappa(F_r(\check{U})) = F_r(\check{U})$ (see (1.25). The material below analyzes certain elements in $F_r(\check{U})$ using the map defined in Definition 7.3 and the zonal spherical functions. It turns out that the notation is better if we first twist $\mathbf{L}$ using κ. In particular, we set

$$\mathcal{L} = \kappa \circ \mathbf{L} \circ \kappa^{-1}. \tag{7.5}$$

LEMMA 7.4. *If $a \in F_r(\check{U})$ then $\mathcal{L}(a) \in \check{U}^B$. Moreover, we have:*
(i) $\kappa((\mathrm{ad}_r\ B_+)\check{U}))\tau(\tilde{\rho}) \subseteq B_+\tau(\tilde{\rho})\check{U} + \tau(\tilde{\rho})\check{U}\chi(B_+)$ *and*
(ii) $\mathcal{L}(a)\tau(\tilde{\rho}) \in \mathcal{C}a + (B_+\tau(\tilde{\rho})\check{U} + \tau(\tilde{\rho})\check{U}\chi(B_+))$.

PROOF. Recall that the algebra $\check{U}^B$, which is defined as the $(\mathrm{ad}_r\ B)$ invariants of $\check{U}$, is also equal to the centralizer of B inside of $\check{U}$ (see (1.28)). Since κ restricts to a conjugate linear antiautomorphism of B, it follows that $\check{U}^B = \kappa(\check{U}^B)$. Hence $\mathcal{L}(a) \in \check{U}^B$. The fact that $\kappa(B) = B$ further ensures that $\kappa((B\check{T}_\Theta)_+\check{U}) = \check{U}(B\check{T}_\Theta)_+$. By the discussion preceding Lemma 7.2, we have that $\chi(u) = \tau(\tilde{\rho})(u)\tau(-\tilde{\rho})$ for all $u \in \check{U}$. It follows that $\kappa\chi = \chi^{-1}\kappa$. Hence $\kappa(\chi^{-1}(B)) = \chi(\kappa(B)) = \chi(B)$. Thus

$$\kappa(\chi^{-1}(B_+)\tau(\tilde{\rho})\check{U} + \tau(\tilde{\rho})\check{U}B_+) = B_+\tau(\tilde{\rho})\check{U} + \tau(\tilde{\rho})\check{U}\chi(B_+).$$

On the other hand, Lemma 7.1 and Lemma 7.2 implies that $\tau(\tilde{\rho})((\mathrm{ad}_r\ B_+)\check{U})$ is a subspace of $\chi^{-1}(B_+)\tau(\tilde{\rho})\check{U} + \tau(\tilde{\rho})\check{U}B_+$. Assertion (i) now follows from the fact that $\tau(\tilde{\rho})\kappa(b) = \kappa(b\tau(\tilde{\rho}))$ for all $b \in \check{U}$.

Using the definitions of $\mathbf{L}$ and $\mathcal{L}$ we see that

$$\mathcal{L}(a)\tau(\tilde{\rho}) = \kappa(\tau(\tilde{\rho})\mathbf{L}(\kappa^{-1}(a))) \in \kappa(\kappa^{-1}(a) + (\mathrm{ad}_r\ B_+)a).$$

Assertion (ii) now follows from (i). □

The next lemma focuses on the connection between the maps $\mathcal{L}$ and $\mathcal{P}_B$ and the evaluation of the zonal spherical functions on elements in $F_r(\check{U})$. The definition of the zonal spherical function $g_{2\lambda}$ and its image $\varphi_{2\lambda}$ under the map Υ can be found in Section 1.

LEMMA 7.5. *Suppose there exists $b \in (B\check{T}_\Theta)_+\check{U}$ and $a \in \check{U}^o$ such that $a + b \in F_r(\check{U})$. Then*

$$z^{2\lambda}(\mathcal{P}_B(\mathcal{L}(a+b)))\varphi_{2\lambda}(\tau(\tilde{\rho})) = \varphi_{2\lambda}(a\tau(\tilde{\rho}))$$

for all $\lambda \in P^+(\Sigma)$.

PROOF. By Theorem 5.4 we have that

$$\begin{aligned} g_{2\lambda}(\mathcal{L}(a+b)\tau(\tilde{\rho})) =& z^{2\lambda}(\mathcal{P}_B(\mathcal{L}(a+b)))g_{2\lambda}(\tau(\tilde{\rho})) \\ =& z^{2\lambda}(\mathcal{P}_B(\mathcal{L}(a+b)))\varphi_{2\lambda}(\tau(\tilde{\rho})). \end{aligned}$$

By Lemma 7.4(i) $g_{2\lambda}(c\tau(\tilde{\rho})) = 0$ for all $c \in \kappa((\mathrm{ad}_r\ B_+)\check{U})$. Note further that $g_{2\lambda}(b\tau(\tilde{\rho})) = 0$ since $b\tau(\tilde{\rho}) \in (B\check{T}_\Theta)_+\check{U}$. It follows from Lemma 7.4 (ii) that

$$g_{2\lambda}(\mathcal{L}(a+b)\tau(\tilde{\rho})) = g_{2\lambda}((a+b)\tau(\tilde{\rho})) = \varphi_{2\lambda}(a\tau(\tilde{\rho})).$$

□

Set $\mathcal{A}_{\leq} = \{\tau(-2\mu)|\ 2\mu \in Q^+(\Sigma) \text{ and } \mu \in P(\Sigma)\}$. Recall the notion of "top" used to denote the top homogenous term of an element in $\mathcal{C}[\mathcal{A}]$ with respect to the degree filtration defined by (5.18). The next lemma determines finer information about the image of elements of the form $\mathcal{L}(a)$ under the quantum Harish-Chandra map $\mathcal{P}_B$.

LEMMA 7.6. *Let $\gamma \in P^+(\Sigma)$. Suppose that there exists $b \in (B\check{T}_\Theta)_+\check{U}$ such that $\tau(2\gamma) + b \in F_r(\check{U})$. Then*

$$\mathcal{P}_B(\mathcal{L}(\tau(2\gamma) + b)) \in \tau(2\gamma)\mathcal{A}_{\leq}.$$

Moreover, there exist elements $\beta_1, \ldots, \beta_m$ in $P^+(\Sigma)$ and scalars a_{β_i} such that $\beta_i \leq \gamma$ for each $1 \leq i \leq m$ and

$$\text{top}(\mathcal{P}_B(\mathcal{L}(\tau(2\gamma) + b))) = \sum_{i=1}^{m} a_{\beta_i}\tau(2\beta_i).$$

PROOF. By Definition 7.3, there exists $c \in B_+$ so that $\mathcal{L}(\tau(2\gamma) + b) = \tau(2\gamma) + b + \kappa((\text{ad}_r\ c)(\kappa^{-1}(\tau(2\gamma) + b)))$. By assumption, $\kappa^{-1}(b) \in \check{U}(B\check{T}_\Theta)_+$. Since B is a left coideal, $(\text{ad}_r\ c)(\kappa^{-1}(b)) \in \check{U}(B\check{T}_\Theta)_+$. It follows that

$$\kappa((\text{ad}_r\ c)(\kappa^{-1}(b))) \in (B\check{T}_\Theta)_+\check{U}.$$

Note that $\kappa(\tau(2\gamma)) = \tau(2\gamma)$. Hence

$$\mathcal{P}_B(\mathcal{L}(\tau(2\gamma) + b)) = \mathcal{P}_B(\tau(2\gamma) + \kappa((\text{ad}_r\ c)\tau(2\gamma))).$$

Since $\kappa(F_r(\check{U})) = F_r(\check{U})$, we have $\tau(2\gamma)+\kappa((\text{ad}_r\ c)\tau(2\gamma)) \in (\text{ad}_r\ U)\tau(2\gamma)$. The right adjoint action of the generators of $\check{U}$ (see (1.14)) combined with the defining relations of $\check{U}$ (see Section 1) ensure that

$$(\text{ad}_r\ U)\tau(2\gamma) \subseteq \sum_{\lambda,\mu,\beta} G_\lambda^+ U_{-\mu}^- \tau(2\gamma - 2\beta)$$

where λ, μ, and β are all elements of $Q^+(\pi)$. Note that $G_\lambda^+ = U_\lambda^+\tau(-\lambda)$ and $U_{-\mu}^- = G_{-\mu}^-\tau(-\mu)$. Thus we have that

$$(\text{ad}_r\ U)\tau(2\gamma) \subseteq \sum_{\lambda,\mu,\beta} U_\lambda^+ G_{-\mu}^- \tau(2\gamma - 2\beta - \lambda - \mu)$$

where λ, μ, and β are all elements of $Q^+(\pi)$. It follows from Lemma 4.1(ii), that $(\text{ad}_r\ U)\tau(2\gamma)$ is a subset of

$$B_+\check{U} + \check{U}^0 N_+^+ + \sum_{\gamma,\beta,\lambda \in Q^+(\pi)} \sum_{0 \leq \eta \leq \lambda+\mu} \mathcal{C}\tau(2\gamma - 2\beta - \lambda - \mu + \eta).$$

Now

$$\sum_{\beta,\lambda,\mu \in Q^+(\pi)} \sum_{0 \leq \eta \leq \lambda+\mu} \mathcal{C}\tau(2\gamma - 2\beta - \lambda - \mu + \eta) \subseteq \sum_{\beta \in Q^+(\pi)} \mathcal{C}\tau(2\gamma - \beta).$$

In particular, $\mathcal{P}_B(u) \in \sum_{\beta \in Q^+(\pi)} \mathcal{C}\tau(2\tilde{\gamma} - \tilde{\beta})$ for all $u \in (\text{ad}_r\ U)\tau(2\gamma)$. On the other hand, we know from Theorem 4.6 that $\mathcal{P}_B(u) \in \mathcal{C}[\mathcal{A}]$. This establishes the first assertion. The second assertion now follows immediately from Theorem 5.9. □

The next result provides a criterion for finding the basis for the image of $\check{U}^B$ under the Harish-Chandra map $\mathcal{P}_B$ described in Theorem C. In particular, we show that if $F_r(\check{U})$ contains a set of elements of a particular form, then the image of these elements under $\mathcal{P}_B$ form a basis for $\widetilde{\mathcal{P}_B(\check{U}^B)}$. In this section, we find this nice set of elements in the easiest case, when $P^+(\pi) = P^+(\Sigma)$. The remaining cases are handled in the next two sections.

THEOREM 7.7. *Suppose that for each $\gamma \in P^+(\Sigma)$, there exists $b_\gamma \in (B\check{T}_\Theta)_+\check{U}$ such that $\tau(2\gamma) + b_\gamma \in F_r(\check{U})$. Then*

$$\{\mathcal{P}_B(\mathcal{L}(\tau(2\gamma) + b_\gamma))|\gamma \in P^+(\Sigma)\} \text{ is a basis for } \mathcal{P}_B(\check{U}^B).$$

Moreover, if $\mathcal{P}_B(\check{U}^B)$ is invariant under the dotted action of W_Θ, then $\mathcal{P}_B(\check{U}^B) = \mathcal{C}[\mathcal{A}]^{W_\Theta\circ}$.

PROOF. Suppose that

$$a = \sum_{\beta \in P^+(\Sigma)} a_\beta \mathcal{L}(\tau(2\beta) + b_\beta)$$

is a linear combination over $\mathcal{C}$ of elements in the set

$$\{\mathcal{L}(\tau(2\gamma) + b_\gamma)|\gamma \in P^+(\Sigma)\}.$$

Assume further that at least one of the coefficients a_β is nonzero. We show that $\mathcal{P}_B(a)$ is nonzero using an argument based on [L5, Lemma 7.5]. This establishes the linear independence of the set

$$\{\mathcal{P}_B(\mathcal{L}(\tau(2\gamma) + b_\gamma))|\gamma \in P^+(\Sigma)\} \tag{7.6}$$

over $\mathcal{C}$. By Lemma 7.5, we have that

$$z^{2\lambda}(\mathcal{P}_B(a))\varphi_{2\lambda}(\tau(\tilde{\rho})) = \varphi_{2\lambda}(\sum_{\beta \in P^+(\Sigma)} a_\beta \tau(2\beta + \tilde{\rho})) \tag{7.7}$$

for all $\lambda \in P^+(\Sigma)$. Since $\varphi_{2\lambda}$ is W_Θ invariant, we further have that

$$\varphi_{2\lambda}(\sum_{\beta \in P^+(\Sigma)} a_\beta \tau(2\beta + \tilde{\rho})) = |W_\Theta|^{-1}\varphi_{2\lambda}(\sum_{w \in W_\Theta} \sum_{\beta \in P^+(\Sigma)} a_\beta \tau(2w\beta + w\tilde{\rho})). \tag{7.8}$$

Here we are assuming in the above sum that each β is a dominant integral restricted weight. It follows that $2\beta + \tilde{\rho}$ is in $P^+(\Sigma)$ for each choice of β. Since at least one of the a_β is nonzero, we must have that $\sum_{w \in W_\Theta} \sum_\beta a_\beta \tau(2w\beta + w\tilde{\rho})$ is nonzero. As explained in Section 5 (see the discussion following Lemma 5.1), there is a nondegenerate bilinear pairing between $\mathcal{C}[P(2\Sigma)]^{W_\Theta}$ and $\mathcal{C}[\mathcal{A}]^{W_\Theta}$. Moreover, this pairing is given by $< f, k >= f(k)$ for all $f \in \mathcal{C}[P(2\Sigma)]^{W_\Theta}$ and $k \in \mathcal{C}[\mathcal{A}]^{W_\Theta}$. It follows that the right hand side of (7.8) is nonzero. Hence (7.7) now implies that $z^{2\lambda}(\mathcal{P}_B(a))$ is nonzero and the set (7.6) is linearly independent over $\mathcal{C}$.

By the previous lemma, $\mathcal{P}_B(\mathcal{L}(\tau(2\gamma) + b_\gamma)) \in \tau(2\gamma)\mathcal{A}_\leq$. Let

$$S_\gamma = \text{span}\{\mathcal{P}_B(\mathcal{L}(\tau(2\beta) + b_\beta))|\beta \in P^+(\Sigma) \text{ and } \beta \leq \gamma\}.$$

It follows that the dimension of S_γ over $\mathcal{C}$ is just the cardinality of the set $\{\beta \in P^+(\Sigma) \text{ and } \beta \leq \gamma\}$. On the other hand, Lemma 7.6 ensures that $\{\text{top}(a)|a \in S_\gamma\}$ is a subspace of $\text{span}\{\tau(2\beta)|\beta \in P^+(\Sigma) \text{ and } \beta \leq \gamma\}$. Since the dimension of

$\{\text{top}(\mathcal{P}_B(a))|a \in S_\gamma\}$ is equal to the dimension of S_γ and both are finite dimensional, it follows that

$$\{\text{top}(a)|a \in S_\gamma\} = \text{span}\{\tau(2\beta)|\beta \in P^+(\Sigma) \text{ and } \beta \leq \gamma\}.$$

This forces

$$\text{top}(\mathcal{P}_B(\mathcal{L}(\tau(2\beta) + b_\beta))) = \tau(2\beta) \tag{7.9}$$

up to a nonzero scalar for all $\beta \in P^+(\Sigma)$. Hence Theorem 5.9 and the above discussion yields

$$\begin{aligned}\{\text{top}(\mathcal{P}_B(a))|a \in \check{U}^B\} &\subseteq \text{span}\{\tau(2\beta)|\beta \in P^+(\Sigma)\}\\ &= \text{span}\{\text{top}(\mathcal{P}_B(\mathcal{L}(\tau(2\beta) + b_\beta)))|\beta \in P^+(\Sigma)\}\\ &\subseteq \text{span}\{\text{top}(\mathcal{P}_B(a))|a \in \check{U}^B\}.\end{aligned}$$

The theorem now follows from the description of the "top" of elements in $\mathcal{C}[\mathcal{A}]^{W_{\Theta^\circ}}$ given in (6.6). □

In order to apply Theorem 7.7, we need to show that $F_r(\check{U})$ contains a set of elements

$$\{\tau(2\gamma) + b_\gamma|\ \gamma \in P^+(\Sigma)\} \tag{7.10}$$

such that each $b_\gamma \in (B\check{T}_\Theta)_+\check{U}$. Recall that $F_r(\check{U}) \cap \check{T}$ is equal to $\{\tau(2\mu)|\ \mu \in P^+(\pi)\}$ (see Section 1). The next lemma shows that it is very easy to find b'_γ when $\gamma \in \widetilde{P^+(\pi)}$.

Lemma 7.8. *Suppose that $\gamma \in P^+(\Sigma)$ and $\gamma = \tilde{\beta}$ for some $\beta \in P^+(\pi)$. Then*

$$\tau(2\gamma) + b'_\gamma \in F_r(\check{U})$$

where $b'_\gamma = \tau(2\beta) - \tau(2\gamma)$. Moreover, $b'_\gamma \in \mathcal{C}[\check{T}_\Theta]_+\check{T}$.

Proof. Since $\beta \in P^+(\pi)$, it follows that $\tau(2\beta) \in F_r(\check{U})$. Therefore $\tau(2\gamma) + b'_\gamma = \tau(2\beta)$ is in $F_r(\check{U})$. Since $\gamma = \tilde{\beta} = (\beta - \Theta(\beta))/2$, we have

$$\begin{aligned}b'_\gamma &= \tau(2\beta) - \tau(2\gamma) = \tau(\beta + \Theta(\beta))\tau(2\gamma) - \tau(2\gamma)\\ &= [\tau(\beta + \Theta(\beta)) - 1]\tau(2\gamma).\end{aligned}$$

The lemma follows from the fact that $\tau(\beta + \Theta(\beta)) - 1 \in \mathcal{C}[\check{T}_\Theta]_+$. □

The following result establishes Theorem C for those symmetric pairs with $\widetilde{P^+(\pi)} = P^+(\Sigma)$.

Theorem 7.9. *Suppose $\mathfrak{g}, \mathfrak{g}^\theta$ satisfies $\widetilde{P^+(\pi)} = P^+(\Sigma)$ or $\mathfrak{g}, \mathfrak{g}^\theta$ is of type FII. Then for all $\gamma \in P^+(\Sigma)$, there exists $b_\gamma \in (B\check{T}_\Theta)_+\check{U}$ such that $\tau(2\gamma) + b_\gamma \in F_r(\check{U})$.*

Proof. The theorem follows immediately from the previous lemma when $\widetilde{P^+(\pi)} = P^+(\Sigma)$. Assume $\mathfrak{g}, \mathfrak{g}^\theta$ is of type FII. By Lemma 2.5(i) and Lemma 7.8,

$$\tau(2m\omega'_4) \in F_r(\check{U}) + (\check{B}\check{T}_\Theta)_+\check{U}$$

for all integers $m \geq 2$. Hence it is sufficient to show

$$\tau(2\omega'_4) \in F_r(\check{U}) + (\check{B}\check{T}_\Theta)_+\check{U}.$$

A straightforward computation using the right adjoint action (1.14)

$$(\mathrm{ad}_r\ x_4y_4)\tau(2\omega_4) = ay_4t_4x_4t_4^{-2}\tau(2\omega_4) + b(1-t_4^{-2})\tau(2\omega_4))$$

for some nonzero scalars a and b. On the other hand, it is straightforward to check that $\tilde{\theta}(y_4)t_4^{-1}x_4\tau(2\omega_4)$ is a linear combination of

$$(\mathrm{ad}_r\ x_4x_3x_2x_1x_3x_2x_3x_4)\tau(2\omega_4)$$
$$(\mathrm{ad}_r\ x_4x_4x_3x_2x_1x_3x_2x_3)\tau(2\omega_4)$$

and an element of $\mathcal{M}_+^+\check{U}$. In other words, there exists an element $u \in U$ such that

$$\tilde{\theta}(y_4)t_4x_4t_4^{-2}\tau(2\omega_4) \in (\mathrm{ad}_r\ u)\tau(2\omega_4) + \mathcal{M}_+^+\check{U}.$$

Hence $t_4^{-2}\tau(2\omega_4)$ can be written as a linear combination of $\tau(2\omega_4)$, $(\mathrm{ad}_r\ x_4y_4)\tau(2\omega_4)$, $(\mathrm{ad}_r\ u)\tau(2\omega_4)$, $B_4x_4t_4^{-2}\tau(2\omega_4)$, and an element of $\mathcal{M}_+^+\check{U}$. The theorem now follows from the fact that $-2\tilde{\alpha}_4 + 2\tilde{\omega}_4 = -2\omega_4' + 4\omega_4'$. □

Suppose that we have found $b_\gamma \in (B\check{T}_\Theta)_+\check{U}$ such that $\tau(2\gamma) + b_\gamma \in F_r(\check{U})$ for each $\gamma \in P^+(\Sigma)$. By Lemma 7.5, we have that

$$z^{2\lambda}(\mathcal{P}_B(\mathcal{L}(\tau(2\gamma)+b_\gamma)))\varphi_{2\lambda}(\tau(\tilde{\rho})) = \varphi_{2\lambda}(\tau(2\gamma+\tilde{\rho})) \tag{7.11}$$

for all $\lambda \in P^+(\Sigma)$. By Theorem 5.4, we further have that

$$\varphi_{2\lambda}(\tau(2\gamma+\tilde{\rho})) = \varphi_{2\lambda} * \mathcal{X}(\mathcal{L}(\tau(2\gamma)+b_\gamma))(\tau(\tilde{\rho})) \tag{7.12}$$

for all $\lambda \in P^+(\Sigma)$. Now the compact zonal spherical functions have been identified with Macdonald polynomials for a large class of symmetric pairs (see for example [L5]). It follows from (7.11) and (7.12) that the set $\{\mathcal{X}(\mathcal{L}(\tau(2\gamma)+b_\gamma))|\ \gamma \in P^+(\Sigma)\}$ is a basis for the family of commuting difference operators associated to these Macdonald polynomials described in [K, Theorem 6.6]. Thus Theorem 7.7 combined with Theorem 7.9 and its generalizations to other symmetric pairs, Theorem 8.2 and Theorem 9.1 (see also Theorem C of the introduction), provide a natural quantum interpretation of these difference operators.

For the other types of irreducible symmetric pairs, finding appropriate elements b'_γ in $(B\check{T}_\Theta)_+\check{U}$ so that $\tau(2\gamma)+b'_\gamma$ is in $F_r(\check{U})$ for all $\gamma \in P^+(\Sigma)$ is more delicate. Set t equal to the rank of Σ. Let $\mu_1, \dots, \mu_t$ denote the simple roots for the restricted root system Σ. Let $\eta_1, \dots, \eta_t$ denote the corresponding fundamental weights in $P^+(\Sigma)$. The next lemma reduces the work to finding just a finite number of elements associated to the weights $\eta_1, \dots, \eta_t$.

Lemma 7.10. *Suppose that for each $B \in \mathcal{B}$ there exists a subset $\{b_i^B|1 \le i \le t\}$ of $(B\check{T}_\Theta)_+\check{U}$ such that $\{\tau(2\eta_i) + b_i^B|1 \le i \le t\}$ is a subset of $F_r(\check{U})$. Then for each $B \in \mathcal{B}$, $F_r(\check{U})$ contains a subset of the form (7.10).*

Proof. Let S be the subset of $\{\tau(2\gamma)|\ \gamma \in P^+(\Sigma)\}$ consisting of those $\tau(2\gamma)$ which are contained in $F_r(\check{U}) + (B\check{T}_\Theta)_+\check{U}$ for every $B \in \mathcal{B}$. To prove the lemma, it is sufficient to show that S equals $\{\tau(2\gamma)|\ \gamma \in P^+(\Sigma)\}$. By assumption, S contains the subset $\{\tau(2\eta_i)|\ 1 \le i \le t\}$. Hence, it is sufficient to show that S is multiplicatively closed.

Fix η and β so that $\tau(2\eta)$ and $\tau(2\beta)$ are both in S. The map $u \mapsto \tau(\eta)u\tau(-\eta)$ defines a Hopf algebra automorphism of $\check{U}$, which we denote by ψ. Note that elements of $\check{U}^0$ are fixed under the action of ψ. It follows that ψ permutes the elements of $\mathcal{B}$. Let $B \in \mathcal{B}$ and choose $B'' \in \mathcal{B}$ such that $\psi(B'') = B$. Choose b

in $(B\check{T}_\Theta)_+\check{U}$ and c in $(B''\check{T}_\Theta)_+\check{U}$ so that $\tau(2\eta)+b$ and $\tau(2\beta)+c$ are elements of $F_r(\check{U})$. Note that $\psi(c)$ is in $(B\check{T}_\Theta)_+\check{U}$. Hence

$$\begin{aligned}(\tau(2\eta)+b)(\tau(2\beta)+c) =&\tau(2\eta+2\beta)+b(\tau(2\beta)+c)+\tau(2\eta)c\\ =&\tau(2\eta+2\beta)+b(\tau(2\beta)+c)+\psi(c)\tau(2\eta)\\ \in&\tau(2\eta+2\beta)+(B\check{T}_\Theta)_+\check{U}.\end{aligned}$$

Thus S contains $\tau(2\beta)\tau(2\eta)=\tau(2\eta+2\beta)$. □

CHAPTER 8

Symmetric Pairs Related to Type AII

In this section, we focus on one family of irreducible symmetric pairs closely related to symmetric pairs of type AII. In particular, let B be a fixed coideal subalgebra in the set $\mathcal{B}$ associated to $\mathfrak{g}, \mathfrak{g}^\theta$. Write t for the rank of the restricted root system associated to $\mathfrak{g}, \mathfrak{g}^\theta$. Assume that $\mathfrak{g}$ contains a θ invariant Lie subalgebra $\mathfrak{r}$ such that $\mathfrak{r}, \mathfrak{r}^\theta$ is of type AII and the rank of the restricted root system associated to $\mathfrak{r}, \mathfrak{r}^\theta$ is $t-1$. Assume further that $t \geq 4$ and $\mathfrak{g}$ is of classical type. In this section, we complete the proof of the quantum version of Helgasons's theorem (Theorem C of the introduction). In particular, we prove the following generalization of Theorem 6.5 and Theorem 6.6.

THEOREM 8.1. *Let $\mathfrak{g}, \mathfrak{g}^\theta$ be an irreducible symmetric pair. Then $\mathcal{P}_B(Z(\check{U})) = \mathcal{C}[\mathcal{A}]^{W_{\Theta^\circ}}$ if and only if $\mathfrak{g}, \mathfrak{g}^\theta$ is not of type EIII, EIV, EVII, and EIX. Moreover, in these cases, $\mathcal{P}_B(\check{U}^B) = \mathcal{P}_B(Z(\check{U}))$.*

The other main objective of this section is finding the special set of elements of $F_r(\check{U})$ described in (7.10) for this family of irreducible symmetric pairs. In particular, we prove the following.

THEOREM 8.2. *Suppose $\mathfrak{g}, \mathfrak{g}^\theta$ contains a symmetric pair $\mathfrak{r}, \mathfrak{r}^\theta$ of type AII and the rank of the restricted root system associated to $\mathfrak{r}, \mathfrak{r}^\theta$ is equal to $t-1$ with $t \geq 4$. Then for each $\gamma \in P^+(\Sigma)$, there exists $b_\gamma \in (B\check{T}_\Theta)_+\check{U}$ such that $\tau(2\gamma)+b_\gamma \in F_r(\check{U})$.*

The possible symmetric pairs that satisfy the assumptions presented in the previous paragraph are given in (2.3)-(2.7). The simple restricted roots associated to these pairs are described in Section 2 (see (2.2) and the list (2.3)-(2.7)). In each case, the (positive) simple restricted roots can be written in terms of the ordinary root system as the set $\{\tilde{\alpha}_{2j} |\ 1 \leq j \leq t\}$. In order to make the notation simpler, we set $\mu_j = \tilde{\alpha}_{2j}$ for each j such that $1 \leq j \leq t$. Thus $\{\mu_1, \dots, \mu_t\}$ is the set of simple roots for Σ. It should be noted that the ordering of these simple roots is the same as given in [H] for the type of root system corresponding to Σ. For each i with $1 \leq i \leq t$, set η_i equal to the fundamental weight in $P(\Sigma)$ corresponding to the simple root μ_i. We further set $\eta_0 = 0$ and $\eta_{t+1} = 0$.

Consider the special case when Σ is a root system of type BC_t. The set Σ is the union of two reduced root systems, one of type B_t and the other of type C_t. Moreover, the set of simple roots for the subroot system of type B_t is $\{\mu_1, \dots, \mu_t\}$ while the set of simple roots for the subroot system of type C_t is $\{\mu_1, \dots, \mu_{t-1}, 2\mu_t\}$. Recall the definition of the fundamental weights associated to the simple roots in Σ given at the beginning of Section 2. It follows that when Σ is of tyep BC_t, the weight lattice $P(\Sigma)$ is equal to the weight lattice associated to the subroot system of Σ of type C_t.

Recall that for each $\mu \in P^+(\pi)$ we can associate a central element $z_{2\mu}$ (see the discussion following Theorem 6.2). By [Jo, Section 7.3], $Z(\check{U})$ is a polynomial ring in the n generators $z_{2\omega_1}, \ldots, z_{2\omega_n}$. Since $\mathcal{P}_B$ is an algebra homomorphism (Theorem 4.4), we have that $\mathcal{P}_B(Z(\check{U}))$ is generated by $\mathcal{P}_B(z_{2\omega_1}), \ldots, \mathcal{P}_B(z_{2\omega_n})$. As explained in Section 6, the top homogeneous term of the W_Θ invariant element $\hat{m}(2\tilde{\mu})$ (defined in (6.5)) is given by $\text{top}(\hat{m}(2\tilde{\mu})) = q^{(\tilde{\rho}, 2\tilde{\mu})}\tau(2\tilde{\mu})$. Since "top" is defined using a degree function on $\mathcal{C}[\mathcal{A}]$, we have $\text{top}(ab) = \text{top}(a)\text{top}(b)$ for all a and b in $\mathcal{C}[\mathcal{A}]$. Thus (6.6) implies that $\mathcal{C}[\mathcal{A}]^{W_{\Theta^\circ}}$ is generated by $\hat{m}(2\eta_i), 1 \leq i \leq n$. Therefore, in order to show $\mathcal{P}_B(Z(\check{U}))$ is equal to $\mathcal{C}[\mathcal{A}]^{W_{\Theta^\circ}}$, it is sufficient to express the $\hat{m}(2\eta_i), 1 \leq i \leq n$ as polynomials in the $\mathcal{P}_B(z_{2\omega_i})$, $1 \leq i \leq n$. The first step is to write the $\mathcal{P}_B(z_{2\omega_i})$ as a linear combination of the $\hat{m}(2\gamma)$ for $\gamma \in P^+(\Sigma)$. The next lemma which analyzes dominant integral weights of the form $\eta_r + \eta_s$ is critical in this process.

LEMMA 8.3. *Assume that Σ is of type A_t. Suppose that r and k are positive integers so that $r \leq k$ and $r + k \leq t + 1$. The set of elements of $P^+(\Sigma)$ which are strictly less than $\eta_r + \eta_k$ is $\{\eta_{r-s} + \eta_{k+s} | 1 \leq s \leq r\}$.*

PROOF. It is straightforward to check that

$$\eta_r + \eta_k - \sum_{i=1}^{s} \sum_{j=r-i+1}^{k+i-1} \mu_j = \eta_{r-s} + \eta_{k+s} \tag{8.1}$$

for $1 \leq s \leq r$. Hence $\{\eta_{r-s} + \eta_{k+s} | 1 \leq j \leq r\}$ is a subset of the intersection of $\{\eta_r + \eta_k - \gamma | \gamma \in Q^+(\Sigma) \setminus \{0\}\}$ with $P^+(\Sigma)$.

Suppose that i satisfies $1 \leq i \leq t$. By [H, Section 13.2, Table 1], the coefficient of μ_t in η_i written as a linear combination of simple restricted roots $\mu_1, \ldots, \mu_t$ is $i/(t+1)$. Now suppose that γ is a nonzero element of $Q^+(\Sigma)$ such that $\eta_r + \eta_k - \gamma \in P^+(\Sigma)$. It follows that the coefficient of μ_t in $\eta_r + \eta_k$ written as a linear combination of the simple restricted roots is $(r + k)/(t + 1)$. On the other hand, the fact that $\gamma \in Q^+(\Sigma)$ implies that the coefficient of μ_t in γ is a nonnegative integer. Since $\eta_r + \eta_k - \gamma$ is in $P^+(\Sigma)$, the coefficient of μ_t in $\eta_r + \eta_k - \gamma$ must be nonnegative. Note that $0 < (r+k)/(t+1) \leq 1$. It follows that the coefficient of μ_t in $\eta_r + \eta_k - \gamma$ written as a linear combination of the simple restricted roots must either be $(r + k)/(t + 1)$ or 0. This latter case can only occur if $r + k = t + 1$. Moreover, the only linear combination of the $\mu_1, \mu_2, \ldots, \mu_{t-1}$ contained in $P^+(\Sigma)$ is 0. Hence this latter case happens only when $\eta_r + \eta_k - \gamma = 0 = \eta_0 + \eta_{t+1}$.

Now assume that the coefficient of μ_t in $\eta_t + \eta_k - \gamma$ is $(r + k)/(t + 1)$. Write $\eta_r + \eta_k - \gamma = \sum_{i=1}^{t} m_i \eta_i$ where the m_i are nonnegative integers. Using the previous paragraph, a comparison of the coefficient of μ_t in both sides of this equation yields

$$\sum_i \frac{m_i i}{(t+1)} = \frac{(r+k)}{(t+1)}. \tag{8.2}$$

We do a similar comparison using the restricted root μ_1. By [H, Section 13.2, Table 1], the coefficient of μ_1 in η_i is $(t - i + 1)/(t + 1)$. Hence the coefficient of μ_1 in $\sum_i m_i \eta_i$ written as a linear combination of the μ_i is

$$\sum_i \frac{m_i(t - i + 1)}{(t + 1)} = \sum_i m_i - \sum_i \frac{m_i i}{(t+1)} = \sum_i m_i - \frac{(r+k)}{(t+1)}$$

On the other hand, the coefficient of μ_1 in $\eta_r + \eta_k$ is

$$\frac{(t - r + 1)}{(t + 1)} + \frac{(t - k + 1)}{(t + 1)} = 2 - \frac{(r + k)}{(t + 1)}.$$

The fact that $\eta_r + \eta_k > \sum_i m_i\eta_i$ ensures that

$$0 \leq \sum_i m_i - \frac{(r+k)}{(t+1)} < 2 - \frac{(r+k)}{(t+1)}.$$

Since the m_i are nonnegative integers, there are two possibilities. The first is that there exists an i with $1 \leq i \leq t$, $m_i = 1$, and $m_j = 0$ for $j \neq i$. By (8.2), the coefficient μ_t in $\eta_r + \eta_k - \gamma$ is $(r+k)/(t+1)$. It follows from [H, Section 13.2, Table 1] that $i = r+k$. Thus $\eta_r + \eta_k - \gamma = \eta_{r+k} = \eta_0 + \eta_{r+k}$.

In the second case, there exists i and j with $1 \leq i, j \leq t$, $m_i = m_j = 1$, and $m_k = 0$ for $k \notin \{i, j\}$. Since the coefficient of μ_t in $\eta_r + \eta_k - \gamma$ is $(r+k)/(t+1)$, we must have that $\eta_r + \eta_k - \gamma = \eta_i + \eta_j$ with $i + j = k + r$. If $i \geq r$, then (8.1) implies that $\eta_i + \eta_j \geq \eta_r + \eta_k$. Thus $i < r$. In particular, we can write $i = r - s$ and $j = k + s$ for some $1 \leq s \leq r$. □

Consider the case when $\mathfrak{g}, \mathfrak{g}^\theta$ is an irreducible symmetric pair of type AII. It follows that the restricted root system is of type A_t and the rank n of $\mathfrak{g}$, is equal to $2t+1$. For this particular irreducible symmetric pair, it is straightforward to check that the only elements of $Q(\pi)$ fixed by Θ are elements of $Q(\pi_\Theta)$. It follows that for each $\alpha \in Q(\pi)$, $\tilde{\alpha} = 0$ if and only if $\alpha \in Q(\pi_\Theta)$.

Given a root α in the root system Δ generated by π, let s_α denote the reflection associated to the root α in the big Weyl group W. The next lemma determines the number of weights in the W orbit of ω_{i+k} whose restriction is equal to $\eta_{i-s} + \eta_{k+s}$ where $k = i$ or $k = i+1$.

LEMMA 8.4. *Assume that $\mathfrak{g}, \mathfrak{g}^\theta$ is an irreducible symmetric pair of type AII. Suppose i is an integer such that $1 \leq i \leq (t+1)/2$. For all integers s such that $0 \leq s \leq i$, the number of elements in the set*

$$\{\beta \in W\omega_{2i} | \tilde{\beta} = \eta_{i-s} + \eta_{i+s}\}$$

equals 2^s. Similarly, suppose $0 \leq i \leq t/2$. Then for all s such that $0 \leq s \leq i$, the number of of elements in the set

$$\{\beta \in W\omega_{2i+1} | \tilde{\beta} = \eta_{i-s} + \eta_{i+1+s}\}$$

equals 2^{s+1}.

PROOF. Fix i such that $1 \leq i \leq (t+1)/2$. Set $\gamma_0 = \omega_{2i}$. For each s such that $1 \leq s \leq i$, set

$$\gamma_s = \omega_{2i-2s+1} - \omega_{2i-2s+2} + \omega_{2i-2s+3} + \cdots - \omega_{2i-2s+(4s-2)} + \omega_{2i-2s+(4s-1)}.$$

Recall that $\mu_j = \tilde{\alpha}_{2j}$ for $1 \leq j \leq t$. Hence by (2.2), if $2j+1$ is not an element of the set $\{2i-2s+1, 2i-2s+2, 2i+2s-2, 2i+2s-1\}$, then $(\gamma_s, \mu_j) = 0$. Since j is an integer, $2j+1$ cannot equal $2i-2s+2$ or $2i+2s-2$. Temporarily set $\mu_0 = \mu_{t+1} = 0$. If $2j+1 = 2i-2s+1$, then $j = i-s$ and $(\gamma_s, \mu_j) = (\omega_{2i-s}, \mu_j) = (\eta_j, \mu_j)$. Similarly, if $2j-1 = 2i+2s-1$ then $j = i+s$ and $(\gamma_s, \mu_j) = (\omega_{2i+s}, \mu_j) = (\eta_j, \mu_j)$. Thus $\tilde{\gamma}_s = \eta_{i-s} + \eta_{i+s}$. (Note that these computations work even in the special case when $i = s$.)

We claim that $\gamma_s \in W\omega_{2i}$ for $0 \leq s \leq i$. Now $\gamma_0 = \omega_{2i}$ and so $\gamma_0 \in W\omega_{2i}$. Also, $(\gamma_0, \alpha_{2i}) = 1$ and in particular, $s_{\alpha_{2i}}\gamma_0 = \gamma_0 - \alpha_{2i} = \omega_{2i-1} - \omega_{2i} + \omega_{2i+1} = \gamma_1$. Hence $\gamma_1 \in W\omega_{2i}$. Now assume that $\gamma_s \in W\omega_{2i}$. It is straightforward to check that

$$\gamma_{s+1} = \gamma_s - \alpha$$

where

$$\alpha = \alpha_{2i-2s-1} + \alpha_{2i-2s} + \alpha_{2i-2s+1} + \cdots + \alpha_{2i+2s+1}.$$

Note that α is a positive root in Δ. Moreover, $(\gamma_s, \alpha) = 1$. Hence $s_\alpha \gamma_s = \gamma_{s+1}$. Therefore, the claim follows by induction on s.

Note that the set

$$\{\beta \in W\omega_{2i} | \tilde{\beta} = \eta_{i-s} + \eta_{i+s}\} \tag{8.3}$$

equals the set

$$\{\beta \in W\omega_{2i} | \tilde{\beta} = \tilde{\gamma}_s\}.$$

By the discussion preceding the theorem, $\tilde{\beta} = \tilde{\gamma}_s$ if and only if $\beta - \gamma_s \in Q(\pi_\Theta)$. Recall that W' is the Weyl group associated to the root system generated by π_Θ. Since $\beta \in W\gamma_s$ and $\gamma_s - \beta \in Q(\pi_\Theta)$, it follows that $\beta \in W'\gamma_s$. Hence the number of elements in (8.3) is equal to $|W'|/|\mathrm{Stab}_{W'}\gamma_s|$.

Set $s_i = s_{\alpha_i}$ for $\alpha_i \in \pi$. Note that W' is generated by $\{s_{2r+1} | 0 \leq r \leq t\}$. Morever, each s_{2j+1} has order 2 while $s_{2j+1}s_{2r+1} = s_{2r+1}s_{2j+1}$ for $r \neq j$. Hence W' is isomorphic to $t+1$ copies of $\mathbf{Z}_2$. Thus $|W'| = 2^{t+1}$. On the other hand, $\mathrm{Stab}_{W'}\gamma_s = \{s_{2r+1} | r \neq 2i + 2j + 1 \text{ for } -s \leq j \leq s-1\}$. It follows that $\mathrm{Stab}_{W'}\gamma_s$ is isomorphic to $t + 1 - (2s)$ copies of $\mathbf{Z}_2$. Hence $|\mathrm{Stab}_{W'}\gamma_s| = 2^{t+1-2s}$ and $|W'|/|\mathrm{Stab}_{W'}\gamma_s| = 2^{2s}$. This completes the proof of the first assertion. The second assertion follows in a similar fashion. □

We continue the analysis of the case when $\mathfrak{g}, \mathfrak{g}^\theta$ is of type AII. Note that $\mathfrak{g}, \mathfrak{g}^\theta$ of type AII implies that $\mathfrak{g}$ is of type A_n. Checking the list of fundamental weights in [H, Section 13.2], it is easy to see that for each $1 \leq i \leq n$, there does not exist $\gamma \in P^+(\pi)$ such that $\omega_i > \gamma$. By Lemma 6.3, there exist nonzero scalars a_i such that $\mathcal{P}(a_i z_{2\omega_i})$ is just the sum

$$\sum_{\mu \in W\omega_i} q^{(\rho, 2\mu)} \tau(2\mu).$$

By the discussion following the proof of Lemma 6.3, $\mathcal{P}_B(a_i z_{2\omega_i})$ is a linear combination of the $\tau(2\tilde{\mu})$ with $\tilde{\mu} \in P(\Sigma)$. Moreover, the coefficients are Laurent polynomials in q with nonnegative integer coefficients and the coefficient of $\tau(2\tilde{\omega}_i)$ is nonzero. Theorem 6.2 ensures that we can also write $\mathcal{P}_B(a_i z_{2\omega_i})$ as a linear combination of the $\hat{m}(2\tilde{\mu})$ with $\tilde{\mu} \in P^+(\Sigma)$. The next lemma gives the evaluation at $q = 1$ of the coefficients of the $\hat{m}(2\tilde{\mu})$ in these sums for some of the $\mathcal{P}_B(a_i z_{2\omega_i})$.

LEMMA 8.5. *Assume that $\mathfrak{g}, \mathfrak{g}^\theta$ is an irreducible symmetric pair of type AII. Then for each integer i such that $2 \leq 2i \leq t+1$, there exist Laurent polynomials $f_{2s}(q), 0 \leq s \leq i$, such that $\mathcal{P}_B(z_{2\omega_{2i}})$ is a nonzero scalar multiple of*

$$\sum_{0 \leq s \leq i} f_{2s}(q) \hat{m}(2\eta_{i-s} + 2\eta_{i+s}).$$

Similarly, for each integer i such that $1 \leq 2i+1 \leq t+1$, there exist Laurent polynomials $f_{2s+1}(q), 0 \leq s \leq i$, such that $\mathcal{P}_B(z_{2\omega_{2i+1}})$ is a nonzero scalar multiple of

$$\sum_{0 \leq s \leq i} f_{2s+1}(q) \hat{m}(2\eta_{i-s} + 2\eta_{i+1+s}).$$

Moreover, the coefficients of powers of q in each $f_j(q)$ are nonnegative integers and $f_j(1) = 2^j$ for all $0 \leq j \leq 2i+1$.

PROOF. Let i be a positive integer such that $i \leq (t+1)/2$. By Lemma 8.3, the only restricted dominant integral weights less than or equal to $\tilde{\omega}_{2i}$ are of the form $\eta_{i-s} + \eta_{i+s}$ for $0 \leq s \leq i$. Hence, by the discussion preceding the lemma, for each s such that $0 \leq s \leq i$, there exists a Laurent polynomial $f_{2s}(q)$ in q such that

$$\mathcal{P}_B(a_i z_{2\omega_{2i}}) = \sum_{0 \leq s \leq i} f_{2s}(q)\hat{m}(2\eta_{i-s} + 2\eta_{i+s}).$$

Recall that each W_Θ orbit of a single element in $P(\Sigma)$ contains a single dominant integral weight in $P^+(\Sigma)$. Hence if we expand out $\mathcal{P}_B(a_i z_{2\omega_{2i}})$ as a sum of terms of the form $\tau(2\tilde{\mu})$ with $\tilde{\mu} \in P(\Sigma)$, it follows that the coefficient of $\tau(2\eta_{i-s} + 2\eta_{i+s})$ is $f_{2s}(q)$. On the other hand, this coefficient can be computed using the expression of $\mathcal{P}(a_i z_{2\omega_{2i}})$ in the paragraph preceding the lemma. In particular, $f_{2s}(q)$ is just the sum

$$\sum_{\{\tilde{\gamma} \in W\omega_{2i} | \tilde{\gamma} = \eta_{i-s} + \eta_{i+s}\}} q^{(\rho, 2\tilde{\gamma})}.$$

This shows that each $f_{2s}(q)$ is a Laurent polynomial in q with nonnegative coefficients. Moreover, $f_{2s}(1)$ is just the number of elements γ in $W\omega_{2i}$ such that $\tilde{\gamma} = \eta_{i-s} + \eta_{i+s}$. By Lemma 8.4, we have $f_{2s}(1) = 2^{2s}$. This completes the proof for the central element $z_{2\omega_{2i}}$. The proof for $z_{2\omega_{2i+1}}$ follows in a similar fashion. □

Note that the map which sends $q^{(\rho, 2\eta_i)}\tau(2\eta_i)$ to z^{η_i} for each $1 \leq i \leq t$ defines an algebra isomorphism from $\mathcal{C}[\mathcal{A}]$ onto $\mathcal{C}[P(\Sigma)]$. For each $\lambda \in P(\Sigma)$, set

$$m(\lambda) = \sum_{\gamma \in W_\Theta \lambda} z^\gamma.$$

The image of $\hat{m}(2\lambda)$ under this isomorphism is just $m(\lambda)$. Note that this isomorphism can be extended to an algebra isomorphism of $\mathcal{C}[\tau(\gamma)|\ \gamma \in \mathbf{Q}\Sigma]$ onto $\mathcal{C}[\mathbf{Q}\Sigma]$. Similarly, the definitions of $m(\lambda)$ and $\hat{m}(2\lambda)$ can be extended to elements λ in $\mathbf{Q}\Sigma$. In the next few lemmas, the computations are done using $m(\lambda)$ instead of $\hat{m}(2\lambda)$ in order to make the notation easier.

LEMMA 8.6. *Assume that $\mathfrak{g}, \mathfrak{g}^\theta$ is an irreducible symmetric pair of type AII. Let r and k be two positive integers such that $r \leq k$ and $r + k = t + 1$. Then for all $0 \leq j \leq r$, the coefficient of 1 in $\hat{m}(2\omega_{r-j})\hat{m}(2\omega_{k+j})$ written as a linear combination of the set $\{\hat{m}(2\lambda)|\lambda \in P^+(\Sigma)\}$ is $\binom{t+1}{r-j}$.*

PROOF. Using the isomorphism described above, we replace each $\hat{m}(2\lambda)$ with $m(\lambda)$ in the proof of the lemma. Fix j such that $0 \leq j \leq r$. Note that if $j = r$, then $m(\eta_{r-j})m(\eta_{k+j}) = m(\eta_0)m(\eta_{t+1}) = 1$. Since $\binom{t+1}{0} = 1$, the lemma follows in this case. Hence we may assume that $j < r$.

Note that

$$m(\eta_{r-j}) \in z^{\eta_{r-j}} + \sum_{\gamma < \eta_{r-j}} \mathbf{N} z^\gamma$$

and

$$m(\eta_{k+j}) \in z^{\eta_{k+j}} + \sum_{\gamma < \eta_{k+j}} \mathbf{N} z^\gamma.$$

Hence

$$m(\eta_{r-j})m(\eta_{k+j}) \in z^{\eta_{r-j} + \eta_{k+j}} + \sum_{\gamma < \eta_{r-j} + \eta_{k+j}} \mathbf{N} z^\gamma.$$

Now $m(\eta_{r-j})m(\eta_{k+j})$ is W_Θ invariant. Hence Lemma 8.3 implies that there exists a nonnegative integer a such that

$$m(\eta_{r-j})m(\eta_{k+j}) \in m(\eta_{r-j}+\eta_{k+j}) + am(0) + \sum_{1\leq b<r-j} \mathbf{N}m(\eta_{r-j-b}+\eta_{k+j+b}).$$

Furthermore, $m(0) = z^0 = 1$ and a is the coefficient of 1 in $m(\eta_{r-j})m(\eta_{k+j})$.

Since the root system Σ is of type A_t and $r+k=t+1$, Σ admits a diagram automorphism d which sends μ_{r-j} to μ_{k+j} for $0\leq j\leq r-1$. Let w_0' denote the longest element of the Weyl group W_Θ and note that $w_0' = -d$. We have the following equality of sets

$$\{w\mu_{r-j}|w\in W_\Theta\} = \{ww_0'\mu_{r-j}|w\in W_\Theta\} = \{-w\mu_{k+j}|w\in W_\Theta\}.$$

Thus if we write

$$m(\mu_{r-j}) = \sum_{1\leq i\leq s} z^{\gamma_i},$$

then

$$m(\mu_{k+j}) = \sum_{1\leq i\leq s} z^{-\gamma_i}.$$

It follows that $s=a$. Moreover, s is just the number of elements in the orbit $W_\Theta\mu_{r-j}$. Hence $s = |W_\Theta|/|\mathrm{Stab}_{W_\Theta}\mu_{r-j}|$. Now $\mathrm{Stab}_{W_\Theta}\mu_{r-j}$ is just the subgroup of W_Θ generated by the reflections corresponding to the simple roots μ_k with $k\neq r-j$. In particular, $\mathrm{Stab}_{W_\Theta}\mu_{r-j}$ is isomorphic to the direct product of two groups $W_1\times W_2$ where W_1 is the Weyl group associated to a root system of type A_{r-j-1} and W_2 is the Weyl group associated to a root system of type A_{t-r+j}. Hence, by [H, Section 12.2, Table 1] it follows that

$$a = |W_\Theta||\mathrm{Stab}_{W_\Theta}\mu_{r-j}|^{-1} = \frac{(t+1)!}{(r-j)!(t+1-r+j)!} = \binom{t+1}{r-j}.$$

□

Set $\Sigma_s = \{\mu_1,\ldots,\mu_t\}$, the set of simple resticted roots associated to Σ. Suppose that Σ_s' is a subset of Σ_s. We define here notation associated to Σ_s' which is used in the rest of the section for different choices of such subsets of Σ_s. Let Σ' be the sub root system of Σ generated by Σ_s'. Let W_Θ' denote the Weyl group of Σ'. For each $\lambda\in P(\Sigma)$, let λ' be the element of $P^+(\Sigma')$ such that $(\lambda-\lambda',\gamma)=0$ for all $\gamma\in Q(\Sigma')$. Set

$$\mathcal{N} = \sum_{\mu_i\in\Sigma_s\setminus\Sigma_s'}\sum_{\gamma\in Q^+(\Sigma)} \mathbf{N}z^{-\mu_i-\gamma}.$$

Given a dominant weight λ in $\mathbf{Q}\Sigma$, set

$$m'(\lambda) = \sum_{\gamma\in W_\Theta'\lambda} z^\gamma. \tag{8.4}$$

Note that for $\lambda\in P^+(\Sigma)$, we have

$$m(\lambda)\in z^{\lambda-\lambda'}m'(\lambda') + z^\lambda\mathcal{N}. \tag{8.5}$$

Moreover,

$$m(\lambda)m(\beta)\in z^{\lambda-\lambda'+\beta-\beta'}m'(\lambda')m'(\beta') + z^{\lambda+\beta}\mathcal{N}. \tag{8.6}$$

Consider for the moment the special case when Σ'_s is the empty set. Expression (8.6) becomes

$$m(\lambda)m(\beta) \in z^{\lambda+\beta} + z^{\lambda+\beta} \sum_{\gamma \in Q^+(\Sigma)\setminus\{0\}} \mathbf{N} z^{-\gamma}.$$

Now suppose that $\lambda = \eta_r$ and $\beta = \eta_k$. It follows from Lemma 8.3 that $m(\eta_r)m(\eta_k)$ can be written as a linear combination of elements in the set $\{m(\eta_{r-s} + \eta_{k+s}) |\ 0 \leq s \leq r\}$.

LEMMA 8.7. *Assume that $\mathfrak{g}, \mathfrak{g}^\theta$ is an irreducible symmetric pair of type AII. Let r and k be two positive integers such that $r \leq k$ and $r + k \leq t+1$. Then*

$$\hat{m}(2\eta_r)\hat{m}(2\eta_k) = \sum_{0 \leq s \leq r} \binom{k-r+2s}{s} \hat{m}(2\eta_{r-s} + 2\eta_{k+s}).$$

PROOF. Once again, we use $m(\lambda)$ instead of $\hat{m}(2\lambda)$. First consider the special case when $r = 0$ and $k = t+1$. Note that this forces $s = 0$. Moreover, $m(\eta_0)m(\eta_{n+1}) = m(0)m(0) = 1$ while $\binom{t+1}{0} = 1$. So the lemma follows in this case. Now assume that r and s are chosen so that $0 \leq s \leq r$ and either $r > 0$ or $k < t+1$.

Set $\Sigma'_s = \{\mu_i |\ r-s+1 \leq i \leq k+s-1\}$ and note that Σ'_s is the set of simple roots for a root system of type $A_{k-r+2s-1}$. For each i, the weight η'_i is just the fundamental weight corresponding to the simple root μ_i in the weight lattice associated to the root system Σ'_s. Checking the formulas for the fundamental weights in [H, Section 13.2, Table 1] yields that $\eta'_r + \eta'_k \in Q^+(\Sigma')$. Since $Q^+(\Sigma')$ is a subset of $Q^+(\Sigma)$, it further follows that $\eta'_r + \eta'_k \in Q^+(\Sigma)$. Now $(\mu_i, \mu_b) \leq 0$ for all $\mu_i \in \Sigma'$ and $\mu_b \in \Sigma \setminus \Sigma'$. Thus $(-\eta'_r - \eta'_k, \mu_b)$ is a nonnegative integer for all $\mu_b \in \Sigma \setminus \Sigma'$. Moreover, since Σ is a root system of type A_t, $\mu_b \in \Sigma \setminus \Sigma'$ implies that $(-\eta'_r - \eta'_k, \mu_b) = 0$ unless $b = r-s$ or $b = k+s$.

Now consider the weight $\gamma = \eta_r + \eta_k - \eta'_r - \eta'_k$. The above discussion ensures that $(\gamma, \mu_b) = 0$ for all $\mu_b \in \Sigma \setminus \{\mu_{r-s}, \mu_{k+s}\}$ while $(\gamma, \mu_b) \geq 0$ for $\mu_b \in \{\mu_{r-s}, \mu_{k+s}\}$. Thus γ is linear combination of the fundamental weights η_{r-s} and η_{k+s} with nonnegative integer coefficients. By the previous paragraph, $(\eta_r + \eta_k) - \gamma = \eta'_r + \eta'_k$ is in $Q^+(\Sigma)$. Hence it follows from Lemma 8.3 that γ must equal $\eta_{r-s} + \eta_{k+s}$. In particular, we have

$$\eta_r + \eta_k - \eta'_r - \eta'_k = \eta_{r-s} + \eta_{k+s}. \tag{8.7}$$

Thus, applying (8.6) to $m(\lambda)m(\beta)$ yields

$$m(\eta_r)m(\eta_k) \in z^{\eta_{r-s}+\eta_{k+s}} m'(\eta'_r)m'(\eta'_k) + z^{\eta_r+\eta_k}\mathcal{N}.$$

Consider the following two ways to expand the product $m(\eta_r)m(\eta_k)$. The first is as a linear combination of terms of the form z^β for $\beta \in P(\Sigma)$ while the second is as a linear combination of terms of the form $m(\lambda)$ for $\lambda \in P^+(\Sigma)$. Note that the coefficient of $z^{\eta_{r-s}+\eta_{k+s}}$ in $m(\eta_r)m(\eta_k)$ written the first way is the same as the coefficient of $m(\eta_{r-s} + \eta_{k+s})$ in $m(\eta_r)m(\eta_k)$ written the second way. By (8.7), $z^{\eta_{r-s}+\eta_{k+s}}$ is not an element of $z^{\eta_r+\eta_k}\mathcal{N}$. Hence the coefficient of $z^{\eta_{r-s}+\eta_{k+s}}$ in $m(\eta_r)m(\eta_k)$ equals the coefficient of $z^{\eta_{r-s}+\eta_{k+s}}$ in $z^{\eta_{r-s}+\eta_{k+s}} m'(\eta'_r)m'(\eta'_k)$. But this is the same as the coefficient of 1 in $m'(\eta'_r)m'(\eta'_k)$. Recall that the first simple root in Σ'_s is μ_{r-s+1} and so μ_r is the sth simple root. Also, there are $k-r+2s-1$ simple roots in Σ'_s. Thus, by Lemma 8.6, the coefficient of $z^{\eta_{r-s}+\eta_{k+s}}$ in $z^{\eta_{r-s}+\eta_{k+s}} m'(\eta'_r)m'(\eta'_k)$ is $\binom{k-r+2s}{s}$. □

Assume that $\mathfrak{g}, \mathfrak{g}^\theta$ is of type AII. Given an integer l such that $1 \leq 2l \leq t+1$, set

$$\hat{M}_{2l} = \sum_{0 \leq i \leq r} 2^{2i} \hat{m}(2\eta_{l-i} + 2\eta_{l+i}). \tag{8.8}$$

Similarly, given an integer l satisfying $1 \leq 2l+1 \leq t+1$, set

$$\hat{M}_{2l+1} = \sum_{0 \leq i \leq r} 2^{2i+1} \hat{m}(2\eta_{l-i} + 2\eta_{l+1+i}). \tag{8.9}$$

By Lemma 8.5, $\hat{M}_k$ can be obtained from $\mathcal{P}_B(z_{2\omega_k})$ by writing the latter as a linear combination of the $\hat{m}(2\lambda)$, for $\lambda \in P^+(\Sigma)$, and evaluating each coefficient at $q = 1$. Let M_k be the image of $\hat{M}_k$ in $\mathcal{C}[P(\Sigma)]$ obtained using the isomorphism described before Lemma 8.6.

LEMMA 8.8. *Assume that Σ is of type A_t.*

(i) *Let l be a positive integer such that $1 \leq 2l \leq t+1$. Then $\hat{m}(\eta_{2l})$ is in the span of the set $\{\mathcal{P}_B(z_{2\omega_{2l}})\} \cup \{\hat{m}(2\eta_{l-j})\hat{m}(2\eta_{l+j}) |\ 0 \leq j < l\}$.*
(ii) *Let l be a positive integer such that $1 \leq 2l+1 \leq t+1$. Then $\hat{m}(2\eta_{2l+1})$ is in the span of the set $\{\mathcal{P}_B(z_{2\omega_{2l+1}})\} \cup \{\hat{m}(2\eta_{l-j})\hat{m}(2\eta_{l+1+j}) |\ 0 \leq j < l\}$.*

PROOF. We prove (i). Assertion (ii) follows using a similar argument. We first prove that $m(\eta_{2l})$ is in the $\mathbf{Q}$ span of the set $\{M_{2l}\} \cup \{m(\eta_{l-j})m(\eta_{l+j}) |\ 0 \leq j < l\}$. Recall that

$$\binom{2i}{r} = \binom{2i}{2i-r}$$

for all $0 \leq r \leq 2i$. Hence

$$2^{2i} = (1+1)^{2i} = \sum_{0 \leq r \leq 2i} \binom{2i}{r} = \binom{2i}{i} + 2 \sum_{0 \leq r \leq i-1} \binom{2i}{r}.$$

By Lemma 8.7, it follows that

$$\begin{aligned}
&m(\eta_l)m(\eta_l) + \sum_{1 \leq j \leq l-1} 2m(\eta_{l-j})m(\eta_{l+j}) \\
&= \sum_{0 \leq s \leq l} \binom{2s}{s} m(\eta_{l-s} + \eta_{l+s}) + \sum_{1 \leq j \leq l-1} \sum_{0 \leq s \leq l-j} 2\binom{2j+2s}{s} m(\eta_{l-j-s} + \eta_{l+j+s}) \\
&= \sum_{0 \leq s \leq l} \binom{2s}{s} m(\eta_{l-s} + \eta_{l+s}) + \sum_{1 \leq j \leq l-1} \sum_{j \leq i \leq l} 2\binom{2i}{i-j} m(\eta_{l-i} + \eta_{l+i}).
\end{aligned}$$

One checks that the subset $\{(j,i)|1 \leq j \leq l-1 \text{ and } j \leq i \leq l\} \cup \{(l,l)\}$ of $\mathbf{Z} \times \mathbf{Z}$ is equal to the subset $\{(j,i)|1 \leq i \leq l \text{ and } 1 \leq j \leq i\}$. Hence

$$m(\eta_l)m(\eta_l) + \sum_{1 \leq j \leq l-1} 2m(\eta_{l-j})m(\eta_{l+j})$$

$$= \sum_{0 \leq s \leq l} \binom{2s}{s} m(\eta_{l-s} + \eta_{l+s}) + \sum_{1 \leq i \leq l} \sum_{1 \leq j \leq i} 2\binom{2i}{i-j} m(\eta_{l-i} + \eta_{l+i}) - 2\binom{2l}{0} m(\eta_{2l})$$

$$= \sum_{0 \leq s \leq l} \left(\binom{2s}{s} + \sum_{1 \leq j \leq s} 2\binom{2s}{s-j} \right) m(\eta_{l-s} + \eta_{l+s}) - 2m(\eta_{2l})$$

$$= \sum_{0 \leq s \leq l} 2^{2s} m(\eta_{l-s} + \eta_{l+s}) - 2m(\eta_{2l}) = M_{2l} - 2m(\eta_{2l})$$

Define X by

$$X = \mathcal{P}_B(z_{2\omega_{2l}}) - \hat{m}(2\eta_l)\hat{m}(2\eta_l) - \sum_{1 \leq j \leq l-1} 2\hat{m}(2\eta_{l-j})\hat{m}(2\eta_{l+j}).$$

By Lemma 8.5 and Lemma 8.7 we can write $X - 2\hat{m}(2\eta_{2l})$ as a linear combination of elements in the set $\{\hat{m}(2\eta_{l-i} + 2\eta_{l+i})|0 \leq i \leq l\}$. Set N equal to the $\mathbf{Q}$ vector space $\sum_{0 \leq i \leq l} \mathbf{Q}\hat{m}(\eta_{l-i} + \eta_{l+i})$. The above computations ensure that the coefficient of each $\hat{m}(2\eta_{l-i} + 2\eta_{l+i})$ in X is an element of $(q-1)\mathbf{Q}[q, q^{-1}]$. Now Lemma 8.7 implies that the set

$$\{\hat{m}(2\eta_{l-i})\hat{m}(2\eta_{l+j}) + \mathbf{Q}\hat{m}(2\eta_{2l})|0 \leq i < l\}$$

is a basis for the $\mathbf{Q}$ vector space $N/(\mathbf{Q}\hat{m}(2\eta_{2l}))$. Hence there exist Laurent polynomials $g_0, \ldots, g_l$ in $\mathbf{Q}[q, q^{-1}]$ so that

$$X - 2\hat{m}(2\eta_{2l}) = \sum_{0 \leq i \leq l-1} (q-1)g_i \hat{m}(2\eta_{l-i})\hat{m}(2\eta_{l+j}) + (q-1)g_l \hat{m}(2\eta_{2l}).$$

Thus

$$X - \sum_{0 \leq i \leq l-1} (q-1)g_i \hat{m}(2\eta_{l-i})\hat{m}(2\eta_{l+j}) = (2 + (q-1)g_l)\hat{m}(2\eta_{2l}).$$

The lemma now follows from the fact that $2 + (q-1)g_l$ cannot be zero. $\square$

We use Lemma 8.8 to show that a certain set generates $\mathcal{C}[\mathcal{A}]^{W_\Theta\circ}$ when $\mathfrak{g}, \mathfrak{g}^\theta$ is of type AII.

THEOREM 8.9. *Assume that $\mathfrak{g}, \mathfrak{g}^\theta$ is of type AII. Then $\mathcal{P}_B(\check{U})) = \mathcal{C}[\mathcal{A}]^{W_\Theta\circ}$.*

PROOF. By Theorem 6.2, we have that $\mathcal{P}_B(Z(\check{U}))$ is a subset of $\mathcal{C}[\mathcal{A}]^{W_\Theta\circ}$. Set $R = \mathcal{P}_B(Z(\check{U}))$ and recall that R is generated by the set $\{\mathcal{P}_B(z_{2\omega_i})|1 \leq i \leq n\}$. It is sufficient to show that $\hat{m}(2\eta_j) \in R$ for all $1 \leq j \leq n$. We do this by induction on j. It follows from Lemma 2.4(i) that $\tilde{\omega}_1 = \eta_1$. Hence Lemma 8.3 and Lemma 8.5 ensure that $\mathcal{P}_B(z_{2\omega_1})$ is a nonzero scalar multiple of $\hat{m}(2\eta_1)$. Therefore, $\hat{m}(2\eta_1)$ is in R. Now assume that $\hat{m}(2\eta_k) \in R$ for all $1 \leq k < j$. Assume first that j is even and write $j = 2l$. By the inductive hypothesis, $\hat{m}(2\eta_{l-i})$ and $\hat{m}(2\eta_{l+i})$ are both in R for $1 \leq i < l$. Hence R contains $\hat{m}(2\eta_{l-i})\hat{m}(2\eta_{l+i})$ for $1 \leq i < l$. Now R also contains $\mathcal{P}_B(z_{2\omega_{2l}})$. Thus by Lemma 8.8(i), R contains $\hat{m}(2\eta_{2l})$. The case for j odd is similar using Lemma 8.8(ii) instead of Lemma 8.8(i). $\square$

We now turn our attention to case where $\mathfrak{g}, \mathfrak{g}^\theta$ is not of type AII. It follows from Araki's classification [A] and the list in Section 2 ((2.3)-(2.7)) that Σ is a root system of type BC_t or C_t. Recall that $\mathfrak{g}$ contains a Lie subalgebra $\mathfrak{r}$ of rank $t-1$ such that $\mathfrak{r}, \mathfrak{r}^\theta$ is a symmetric pair of type AII. Note that $\mathfrak{r}$ is a semisimple Lie algebra of type A_{2t-1}. Let $\pi_{\mathfrak{r}}$ denote the subset of simple roots in π which span the root system of $\mathfrak{r}$. Now $\pi_{\mathfrak{r}}$ is just the set $\{\alpha_1, \dots, \alpha_{2t-1}\}$ consisting of the first $2t-1$ elements in π. Moreover, this ordering of the simple roots is the same as the standard ordering coming from the numbering on the corresponding Dynkin diagram given in [H]. Set $W_{\mathfrak{r}}$ equal to the Weyl group of the root system of $\mathfrak{r}$ and note that $W_{\mathfrak{r}}$ is a subgroup of W.

We use below the notation introduced before Lemma 8.7 where Σ_s' equals the set $\{\mu_1, \dots, \mu_{t-1}\}$. Note that $\Sigma_s \setminus \Sigma_s' = \{\mu_t\}$. Hence

$$\mathcal{N} = z^{-\mu_t} \sum_{\gamma \in Q^+(\Sigma)} z^{-\gamma}. \tag{8.10}$$

The next lemma gives technical information which relates certain restricted roots of $\mathfrak{g}, \mathfrak{g}^\theta$ to that of $\mathfrak{r}, \mathfrak{r}^\theta$.

LEMMA 8.10. *Suppose that $\mathfrak{g}, \mathfrak{g}^\theta$ is not of type AII. Then for all i, k, and s such that $1 \leq r \leq k \leq t-1$ and $0 \leq s \leq \min(r, t-k)$ we have*

$$\eta_{r-s} + \eta_{k+s} - \eta'_{r-s} - \eta'_{k+s} = \eta_r + \eta_k - \eta'_r - \eta'_k. \tag{8.11}$$

PROOF. Suppose that r has been chosen so that $1 \leq r \leq t-1$. We have $(\eta_r - \eta'_r, \mu_j) = 0$ for all $\mu_j \in \Sigma'_s$. Since Σ' is of type A_{t-1}, it follows that the coefficient of μ_{t-1} in η'_r is r/t (see [H, Section 13.2]). Hence

$$(\eta_r - \eta'_r, \mu_t) = -r(\mu_{t-1}, \mu_t)/t. \tag{8.12}$$

Assume first that Σ is of type BC_t. It follows that

$$(\mu_{t-1}, 2\mu_t) = -(2\mu_t, 2\mu_t)/2 = -(\eta_t, 2\mu_t).$$

Thus $\eta_r - \eta'_r = r\eta_t/t$. Hence $\eta_r + \eta_k - \eta'_r - \eta'_k = (r+k)\eta_t/t$ for $1 \leq r \leq k \leq t-1$ with $0 \leq s \leq \min(r, t-1-k)$. A similar argument shows that when Σ is of type C_t, $\eta_r - \eta'_r = r\eta_t/t$ for all $1 \leq r \leq t-1$. Now $\eta'_t = 0$ so $\eta_r - \eta'_r = r\eta_t/t$ is also true for $r = t$. It follows that for $1 \leq r \leq k \leq t$ with $0 \leq s \leq \min(r, t-k)$, we have $\eta_r + \eta_k - \eta'_r - \eta'_k = (r+k)\eta_t/t$. Thus $\eta_r + \eta_k - \eta'_r - \eta'_k$ depends only on $r+k$ which proves the lemma. □

Recall that the ordinary Harish-Chandra map $\mathcal{P}$ defines an isomorphism from $Z(\check{U})$ onto $\mathcal{C}[\tau(2\lambda)|\lambda \in P(\pi)]^{W\cdot}$ (see Theorem 3.1). Recall further the definition of $\hat{\tau}(2\gamma)$ given in (6.1) for $\gamma \in P^+(\pi)$. It is straightforward to see that $\hat{\tau}(2\gamma)$ is contained in $\mathcal{C}[\tau(2\lambda)|\lambda \in P(\pi)]^{W\cdot}$. For each i satisfying $1 \leq i \leq n$, let z_i be the element in $Z(\check{U})$ such that $\mathcal{P}(z_i) = \hat{\tau}(2\omega_i)$. The following is an extended version of Lemma 8.5.

LEMMA 8.11. *Suppose that $\mathfrak{g}, \mathfrak{g}^\theta$ is not of type AII. Then for each integer i satisfying $2 \leq 2i \leq t$, there exist Laurent polynomials $g_{2s}(q)$, $0 \leq s \leq i$, such that*

$$\mathcal{P}_B(z_{2i}) = \sum_{0 \leq s \leq i} g_{2s}(q)\hat{m}(2(\eta_{i-s} + \eta_{i+s})) + \sum_{\gamma \in Q^+(\Sigma)} \mathbf{Q}(q)\tau(2(2\eta_i - \mu_t - \gamma))$$

up to a nonzero scalar. Similarly, for each integer i such that $1 \leq 2i+1 \leq t+1$, there exist Laurent polynomials $g_{2s+1}(q)$, $0 \leq s \leq i$ such that

$$\mathcal{P}_B(z_{2i+1}) = \sum_{0\leq s\leq i} g_{2s+1}(q)\hat{m}(2(\eta_{i-s}+\eta_{i+1+s})) + \sum_{\gamma\in Q^+(\Sigma)} \mathbf{Q}(q)\tau(2(\eta_i+\eta_{i+1}-\mu_t-\gamma))$$

up to a nonzero scalar. Moreover, the coefficients of the powers of q in each $g_j(q)$ are nonnegative integers and $g_j(1) = 2^j$ for $0 \leq j \leq 2i+1$.

PROOF. Let ψ denote the isomorphism from $\mathcal{C}[\tau(2\gamma)|\gamma \in P(\pi)]$ to $\mathcal{C}[P(\pi)]$ defined by $\psi(\tau(2\gamma)q^{(\rho,2\gamma)}) = z^\gamma$ for all $\gamma \in P(\pi)$. Note that for each i, $\psi(\mathcal{P}(z_i)) = \sum_{\beta\in W\omega_i} z^\beta$ up to a nonzero scalar. Furthermore,

$$\sum_{\beta\in W\omega_i} z^\beta = \sum_{\beta\in W_{\mathfrak{r}}\omega_i} z^\beta + \sum_{\gamma\in W\omega_i\setminus W_{\mathfrak{r}}\omega_i} z^\gamma. \tag{8.13}$$

Suppose that $\gamma \in W\omega_i \setminus W_{\mathfrak{r}}\omega_i$. It follows that there exists a reflection $s = s_{\alpha_j}$ corresponding to a simple root α_j in $\pi \setminus \pi_{\mathfrak{r}}$ and elements $w_1 \in W_{\mathfrak{r}}$ and $w_2 \in W$ such that $\gamma = w_2 s w_1 \omega_i$. We may assume that $sw_1\omega_i \neq w_1\omega_i$ and thus $(\alpha_j, w_1\omega_i)$ is nonzero. We may further assume that there is a reduced expression for w_2sw_1 equal to the product of the reduced expression for w_2, s, and the reduced expression for w_1 in that order. It follows that

$$\omega_i - \gamma - \alpha_j \in Q^+(\pi). \tag{8.14}$$

Now $(\alpha_j, \omega_i) = 0$ since $\alpha_i \in \pi_{\mathfrak{r}}$. Furthermore, $\omega_i - w_1\omega_i \in Q^+(\pi_{\mathfrak{r}})$ because $w_1 \in W_{\mathfrak{r}}$. Thus (α_j, α) must be nonzero for some $\alpha \in \pi_{\mathfrak{r}}$. We can write $\pi_{\mathfrak{r}} = \{\alpha_1, \ldots, \alpha_k\}$ for some choice of k with $1 \leq k \leq n$. Checking the the list (2.3)-(2.7), one sees that Δ is a root system of type A_n, C_n, or D_n. Hence j must equal $k+1$ and moreover, $\tilde{\alpha}_j$ is the last simple root of Σ. In particular, $k = t-1$ and $\tilde{\alpha}_j = \mu_t$. Hence (8.14) implies that $\tilde{\omega}_i - \mu_t - \tilde{\gamma} \in Q^+(\Sigma)$. Thus (8.13) yields

$$\sum_{\beta\in W\omega_i} z^{\tilde{\beta}} \in \sum_{\beta\in W_{\mathfrak{r}}\omega_i} z^{\tilde{\beta}} + z^{\tilde{\omega}_i}\mathcal{N}.$$

The first assertion of the lemma now follows using (8.5),(8.6), and Lemma 8.5. A similar argument works for the second assertion. □

The next lemma provides additional information about the image of certain central elements under $\mathcal{P}_B$ not covered by Lemma 8.11.

LEMMA 8.12. *The element $\hat{m}(2\eta_1)$ is in $\mathcal{P}_B(Z(\check{U}))$.*

PROOF. Fix i with $1 \leq i \leq t$. Suppose that the only weight in $P^+(\Sigma)$ less than η_i is 0. By the discussion before Lemma 8.5, we can write $\mathcal{P}_B(z_{2\omega_i})$ as a linear combination of the $\hat{m}(2\gamma)$ with $\gamma \in P^+(\Sigma)$ and $\gamma < \eta_i$. Hence there exist scalars a and b with a nonzero such that $\mathcal{P}_B(z_{2\omega_i}) = a\hat{m}(2\eta_i) + b$. It follows that $\hat{m}(2\eta_i) \in \mathcal{P}_B(Z(\check{U}))$.

Since Σ is of type A_t, BC_t, or C_t, it follows by inspection (see [H, Section 13.2]) that there are no elements in $P^+(\Sigma)$ less than η_1. Moreover, Lemma 2.4(i) ensures that $\tilde{\omega}_1 = \eta_1$. Hence, by the previous paragraph, $\hat{m}(2\eta_1) \in \mathcal{P}_B(Z(\check{U}))$. □

For each $1 \leq k \leq n$, let Z_k be the image of $\mathcal{P}_B(z_k)$ in $\mathcal{C}[P(\Sigma)]$ using the isomorphism discussed before Lemma 8.6. Using Lemma 8.11, we see that given integers l, k such that $2 \leq 2l \leq t$ and $1 \leq 2k+1 \leq t$,

$$Z_{2l} \in \sum_{0 \leq s \leq l} g_{2s}(q) m(\eta_{l-s} + \eta_{l+s}) + z^{2\eta_l}\mathcal{N}$$

and

$$Z_{2l+1} \in \sum_{0 \leq s \leq l} g_{2s+1}(q) m(\eta_{k-s} + \eta_{k+1+s}) + z^{\eta_k + \eta_{k+1}}\mathcal{N}$$

where the g_j are Laurent polynomials in q satisfying $g_j(1) = 2^j$. Recall the definition of $m'(\lambda)$ given in (8.4) and set

$$Z'_{2l} = \sum_{0 \leq s \leq l} g_{2s}(q) m'(\eta'_{l-s} + \eta'_{l+s})$$

and

$$Z'_{2l+1} = \sum_{0 \leq s \leq l} g_{2s+1}(q) m'(\eta'_{k-s} + \eta'_{k+1+s}).$$

Assume for the moment that $\mathcal{C}[\mathcal{A}]^{W_\Theta \circ}$ is a subset of $\mathcal{P}_B(Z(\check{U}))$. By Theorem 6.2, it follows that these two algebras are equal. A comparison of top degree terms as in Theorem 6.5, using Theorem 5.9 and (6.6), would then imply that $\mathcal{P}_B(\check{U}^B) = \mathcal{P}_B(Z(\check{U})) = \mathcal{C}[\mathcal{A}]^{W_\Theta \circ}$. Hence the following generalization of Theorem 8.9 completes the proof of Theorem 8.1.

THEOREM 8.13. *Suppose $\mathfrak{g}, \mathfrak{g}^\theta$ contains a symmetric pair $\mathfrak{r}, \mathfrak{r}^\theta$ of type AII and the rank of the restricted root system associated to $\mathfrak{r}, \mathfrak{r}^\theta$ is equal to $t-1$ with $t \geq 4$. Then $\mathcal{C}[\mathcal{A}]^{W_\Theta \circ}$ is a subalgebra of $\mathcal{P}_B(Z(\check{U}))$.*

PROOF. Let R be the image of $\mathcal{P}_B(Z(\check{U}))$ in $\mathcal{C}[P(\Sigma)]$ using the isomorphism described before Lemma 8.6. In particular, R contains Z_k for $1 \leq k \leq n$. It is sufficient to show that $m(\eta_i) \in R$ for $1 \leq i \leq t$. By Theorem 8.9, the result is true when Σ is of type A_t. Hence we may assume that Σ is of type BC_t or C_t. By Lemma 8.12, R contains $m(\eta_1)$. We use induction along the lines of the proof of Theorem 8.9 to show that all the $m(\eta_j)$ are in R.

For each $1 \leq j \leq t-1$, we have that the restriction η'_j of η_j to the root system Σ' is just the fundamental weight in $P^+(\Sigma')$ associated to the root μ_j. By (8.5) and Lemma 8.10, we have that

$$m(\eta_{i-s} + \eta_{k+s}) \in z^{\eta_i + \eta_k - \eta'_i - \eta'_k} m'(\eta'_{i-s} + \eta'_{k+s}) + z^{\eta_i + \eta_k}\mathcal{N}$$

for all i, k, and s such that $1 \leq i \leq k \leq t-1$, $i + k \leq t$, and $0 \leq s \leq i$. It follows from the definition of the Z_k and Z'_k that

$$Z_{2l} \in z^{2\eta_l - 2\eta'_l} Z'_{2l} + z^{2\eta_l}\mathcal{N}$$

for all integers l such that $2 \leq 2l \leq t$ and

$$Z_{2j+1} \in z^{\eta_j + \eta_{j+1} - \eta'_j - \eta'_{j+1}} Z'_{2j+1} + z^{\eta_j + \eta_{j+1}}\mathcal{N}$$

for all integers j such that $1 \leq 2j+1 \leq t$. By (8.6) and Lemma 8.10, we further have that

$$m(\eta_{i-s}) m(\eta_{k+s}) \in z^{\eta_i + \eta_k - \eta'_i - \eta'_k} m'(\eta'_{i-s}) m'(\eta'_{k+s}) + z^{\eta_i + \eta_k}\mathcal{N}$$

for all i, k, and s such that $1 \leq i, k \leq t-1$, $i + k \leq t$, and $0 \leq s \leq i$.

Assume that $m(\eta_1), \ldots, m(\eta_s)$ are in R for some $1 \leq s \leq t-1$. Assume further that

$$s + 1 = 2l \text{ for some integer } l. \tag{8.15}$$

Arguing as in Lemma 8.8, we can write $m'(\eta'_{2l})$ as a linear combination of Z'_{2l} and the elements in the set $\{m'(\eta'_{l-j})m'(\eta'_{l+j}) | 0 \leq j < l\}$. Thus there is an element Y in R which is a linear combination of elements in the set $\{Z_l\} \cup \{m(\eta_{l-j})m(\eta_{l+j}) | 0 \leq j < l\}$ such that

$$Y \in m(\eta_{2l}) + z^{2\eta_l}\mathcal{N}.$$

Consider an element $\gamma \in P^+(\Sigma)$ such that $z^\gamma \in z^{2\eta_l}\mathcal{N}$. By (8.10), we have that $2\eta_l - \mu_t - \gamma \in Q^+(\Sigma)$.

Suppose that Σ is of type BC_t. Recall that the weight lattice $P(\Sigma)$ associated to Σ is equal to the weight lattice associated to the subroot system of type C_t. Moreover, the last simple root of this subroot system is $2\mu_t$. It follows from the list of the fundamental weights in [H, Section 13.2] that the coefficient of μ_t in $2\eta_l$ is l. Now suppose that Σ is of type C_t. Another check using the table in [H, Section 13.2] yields that the coefficient of μ_t in $2\eta_l$ is also l. Hence in either case, the coefficient of μ_t in γ must be a nonnegative integer less than or equal to $l-1$.

Now assume that Σ is of either type, BC_t or C_t. By Lemma 8.12, $m(\eta_1)$ is in R. By (8.15), we have that $2l = s+1 \leq t$. Hence $l - 1 \leq (s-1)/2$. Checking the table in [H, Section 13.2], we see that γ is a linear combination of weights in the set $\{\eta_i |\ 1 \leq i \leq s-1\}$. (Here, we are using the fact that the coefficient of μ_t in γ is less than or equal to $l-1$.) It follows that $Y - m(\eta_{2l})$ is in the subalgebra of R generated by the set $\{m(\eta_i) |\ 1 \leq i \leq s-1\}$. Since $Y \in R$, we must also have $m(\eta_{2l}) \in R$. A similar argument using Lemma 8.8(ii) shows that $m(\eta_{2l+1})$ is in R. Thus by induction, R contains $m(\eta_1), \ldots, m(\eta_t)$. □

We now turn our attention to proving Theorem 8.2. Note that any subset $\{\mu_1, \ldots, \mu_j\}$ of the simple restricted roots Σ_s with j strictly less than t describes a root system of type A_j. Recall the notation introduced before Lemma 8.7 for subsets of Σ_s.

LEMMA 8.14. *Consider j such that $1 \leq j < t$ and let $\Sigma'_s = \{\mu_1, \ldots, \mu_j\}$. Then*

$$\eta'_k + k(j+1)^{-1}\eta_{j+1} = \eta_k$$

for all $1 \leq k \leq j$.

PROOF. Since Σ is a root system of type A_t, BC_t, or C_t, we have that $(\mu_i, \mu_{i+1}) = -(\eta_{i+1}, \mu_{i+1})$ for all $1 \leq i \leq t-1$. Fix k such that $1 \leq k \leq j$. Now $(\eta'_k, \mu_i) = \delta_{ik}(\mu_i, \mu_i)/2$ for all $1 \leq i \leq j$. It follows from [H, Section 13.2, Table 1] that

$$\begin{aligned} \eta'_k = [&(j-k+1)\mu_1 + 2(j-k+1)\mu_2 + \cdots + (k-1)(j-k+1)\mu_{j-1} \\ &+ r(j-k+1)\mu_k + r(j-k)\mu_{k+1} + \cdots + k\mu_j](j+1)^{-1}. \end{aligned} \tag{8.16}$$

Hence $(\eta'_k, \mu_{j+1}) = k(j+1)^{-1}(\mu_j, \mu_{j+1}) = -k(j+1)^{-1}(\eta_j, \mu_{j+1})$. Note that (8.16) also implies that $(\eta'_k, \mu_s) = 0$ for all $s > j+1$. This proves the lemma. □

The proof of Theorem 8.2 involves induction on Lie subalgebras of $\mathfrak{g}$. In particular, given i such that $1 \leq i \leq t-1$, define the subset π_i of π by $\pi_i = \{\alpha_j | 1 \leq j \leq 2i+1\}$. Let $\mathfrak{g}_i$ denote the semisimple Lie subalgebra of $\mathfrak{g}$ generated by the positive and negative root vectors associated to the simple roots in π_i. Note that the root system of $\mathfrak{g}_i$ is of type A_{2i+1}. We have $\mathfrak{g}_i \subseteq \mathfrak{g}_{i+1}$ for $1 \leq i \leq t-2$. Moreover, θ

restricts to an involution on $\mathfrak{g}_i$ such that the symmetric pair $\mathfrak{g}_i, \mathfrak{g}_i^\theta$ is of type AII and the rank of the corresponding restricted root system is i. Given i such that $1 \leq i \leq t-1$, let U_i denote the quantized enveloping algebra of $\mathfrak{g}_i$ considered as a subalgebra of $U_q(\mathfrak{g})$. In particular, U_i is generated by $x_j, y_j, t_j^{\pm 1}$ for all j such that $\alpha_j \in \pi_i$. Set $U_0 = \mathcal{C}$ and $U = U_t$.

Let $\check{U}_i$ be the simply connected quantized enveloping algebra of $\mathfrak{g}_i$. We use $\check{U}_0$ to denote the scalar field $\mathcal{C}$. Note that $P(\pi_i)$ is not necessarily a subset of $P(\pi)$ and so $\check{U}_i$ is not a subring of $\check{U}$. Instead, $\check{U}_i$ is a subring of an extension of $\check{U}$ by elements of the form $\tau(\gamma), \gamma \in P(\pi_i)$. On the other hand, suppose that $\gamma \in P(\pi)$ and $\gamma_i \in \mathbf{Q}\pi_i$ such that $\gamma(\alpha_j) = \gamma_i(\alpha_j)$ for all $\alpha_j \in \pi_i$. Since γ is in $P(\pi)$ it follows that $\gamma_i \in P(\pi_i)$. Consider the ring extension $\check{U}_i\check{T}$ of $\check{U}_i$ by the elements in the group $\check{T}$. Recall that ω_j is the fundamental weight corresponding to the simple root α_j in $P^+(\pi)$. For any $j > 2i+1$ and any rational number r such that $\tau(r\omega_j) \in \check{U}_i\check{T}$, we have that $\tau(r\omega_j)$ is in the center of $\check{U}_i\check{T}$. In particular, there exists a positive integer m such that $\check{U}_i\check{T}$ is isomorphic to a subring of the tensor product algebra $\check{U}_i \otimes_{\mathcal{C}} \mathcal{C}[\tau((1/m)\omega_{2i+2})^{\pm 1}, \dots, \tau((1/m)\omega_n)^{\pm 1}]$. Furthermore, $F_r(\check{U}_i\check{T})$ is a subring of $F_r(\check{U}_i)\mathcal{C}[\tau((1/m)\omega_{2i+2})^{\pm 1}, \dots, \tau((1/m)\omega_n)^{\pm 1}]$.

Recall the map $\mathcal{L}$ defined using Definition 7.3 and (7.5) that maps elements of $F_r(\check{U})$ to elements of $\check{U}^B$. Given i such that $1 \leq i < t$, we define the same type of map, denoted by $\mathcal{L}_i$, for the quantum symmetric pair $U_i, U_i \cap B$. In particular, given $a \notin (\mathrm{ad}_r\ (B \cap U_i)_+)\check{U}_i$, we have that $\mathcal{L}_i(a)$ is equal to $\kappa(b)$ where b is the unique element of $\check{U}_i^{B\cap U_i}$ contained in

$$\kappa^{-1}(a) + (\mathrm{ad}_r\ (B \cap U_i)_+)(\kappa^{-1}(a)). \tag{8.17}$$

Note that $\tau(r\omega_k)$ is in the center of $\check{U}_i\check{T}$ for all k such that $2i+2 \leq k \leq n$ and all rational numbers r with $\tau(r\omega_k) \in \check{U}_i\check{T}$. Thus we can extend $\mathcal{L}_i$ to $\check{U}_i\check{T}$ by insisting that $\mathcal{L}_i(ab) = \mathcal{L}_i(a)b$ for all b that can be expressed as a Laurent polynomial of elements of the form $\tau(r_k\omega_k)$ with r_k rational and $2i+2 \leq k \leq n$. Note that the extended form of $\mathcal{L}_i$ satisfies (8.17) where $\check{U}_i^{B\cap U_i}$ is replaced with $(\check{U}_i\check{T})^{B\cap U_i}$.

Let $\mathcal{P}_{B\cap U_i}$ denote the Harish-Chandra map on $\check{U}_i$ with respect to the quantum symmetric pair $U_i, B \cap U_i$. Note that $\mathcal{P}_B(ab) = \mathcal{P}_{B\cap U_i}(a)b$ for all $a \in \check{U}_i$ and b in the center of $\check{U}_i\check{T}$.

Now suppose that $\Sigma'_s = \{\mu_1, \dots, \mu_j\}$ for some $1 \leq j < t$ and recall the notation introduced before Lemmas 8.7. Given a weight λ in $P^+(\Sigma')$, we set

$$\hat{m}'(2\lambda) = \sum_{\gamma \in W'_\Theta \lambda} q^{(\rho, 2\gamma)}\tau(2\gamma).$$

Note that the image of $\hat{m}'(2\lambda)$ in $\mathcal{C}[\mathbf{Q}\Sigma]$ using the map defined before Lemma 8.6 is just the element $m'(\lambda)$ (see (8.4)).

LEMMA 8.15. *Consider j such that $1 \leq j < t$ and let $\Sigma'_s = \{\mu_1, \dots, \mu_j\}$. Suppose that for each k with $1 \leq k \leq j$, there exists $b_k \in ((B \cap U_j)\check{T}_\Theta)_+\check{U}_j\check{T}$ such that $\{\tau(2\eta_k) + b_k |\ 1 \leq i \leq j\}$ is a subset of $F_r(\check{U})$. Then for all $1 \leq k \leq j$, $\mathcal{L}_j((\tau(2\eta_k) + b_k)$ is an element of $F_r(\check{U})$. Moreover*

$$\mathcal{P}_B(\mathcal{L}_j(\tau(2\eta_k) + b_k)) = \hat{m}'(2\eta'_k)\tau(k(j+1)^{-1}2\eta_{j+1})$$

up to a nonzero scalar.

PROOF. Fix k such that $1 \leq k \leq j$. Recall that $F_r(\check{U})$ is an $(\mathrm{ad}_r\ U)$ module and $\kappa(F_r(\check{U})) = F_r(\check{U})$ (see (1.25)). Hence the definition of $\mathcal{L}_j$ and the fact that $\tau(2\eta_k) + b_k$ is in $F_r(\check{U})$ ensure that $\mathcal{L}_j(\tau(2\eta_k) + b_k)$ is an element of $F_r(\check{U})$. Thus we only need to prove the second assertion of the lemma.

Set $b'_k = b_k\tau(-k(j+1)^{-1}2\eta_{j+1})$. By Lemma 8.14, we have

$$\tau(2\eta'_k + k(j+1)^{-1}2\eta_{j+1}) = \tau(2\eta_k).$$

By Lemma 2.4 (ii), we have that $\tau(2\eta_{j+1}) = \tau(\tilde{\omega}_{2j+2})$. In particular, b'_k is also $B \cap U_j$ invariant. Moreover,

$$\mathcal{L}_j(\tau(2\eta_k) + b_k) = \mathcal{L}_j(\tau(2\eta'_k) + b'_k)\tau(k(j+1)^{-1}2\eta_{j+1}) \tag{8.18}$$

for each $1 \leq k \leq j$.

Since Σ'_s is a root system of type A_j, it follows that there are no weights in $P^+(\Sigma_s)$ less than η'_k. By Lemma 7.6 applied to the symmetric pair $B \cap U_i$, U_i, we see that the top degree term of $\mathcal{P}_{B\cap U_j}(\mathcal{L}_j(\tau(2\eta'_k) + b'_k))$ is a nonzero scalar multiple of $\tau(2\eta'_k)$. Moreover, Lemma 7.6 ensures that $\mathcal{P}_{B\cap U_j}(\mathcal{L}_j(\tau(2\eta'_k) + b'_k))$ is a linear combination of elements of the form $\tau(2\eta'_k - 2\gamma')$ where $\gamma \in Q^+(\Sigma')$. It follows that $\mathcal{P}_{B\cap U_j}(\mathcal{L}_j(\tau(2\eta_k) + b'_k))$ is a nonzero scalar multiple of $\hat{m}'(2\eta'_k)$. The lemma now follows. □

Let j be an integer with $1 \leq j < t$. Write $\omega_{1j}, \ldots, \omega_{2j+1,j}$ for the fundamental weights in the weight lattice $P^+(\pi_j)$ associated to the root system with simple roots π_j. Now π is a root system of type A_n, C_n or D_n, while π_j has type A_{2j+1}. The same argument as in the proof of Lemma 8.14 yields

$$\omega_{kj} + k(2j+2)^{-1}\omega_{2j+2} = \omega_j \tag{8.19}$$

for $1 \leq k \leq 2j+1$.

LEMMA 8.16. *Suppose that j satisfies $1 \leq j < t$ and let $\Sigma'_s = \{\mu_1, \ldots, \mu_j\}$. Then there exist a central element z'_{j+1} in $\check{U}_j$ and a central element z_{j+1} in $\check{U}_j\check{T}$ such that*

(i) $z_{j+1} = z'_{j+1}\tau(\omega_{2j+2})$
(ii) $z_{j+1} \in \check{U}$
(iii) *1 is in the $\mathcal{C}$ span of the set*

$$\{\mathcal{P}_{B\cap\check{U}_j}(z'_{j+1})\} \cup \{\hat{m}'(2\eta'_k)\hat{m}'(2\eta'_{j+1-k})|\ 1 \leq k \leq j\}. \tag{8.20}$$

PROOF. As explained in the beginning of Section 6, there exists a unique central element z'_{j+1} of $\check{U}_j$ contained in

$$\tau(2\omega_{j+1,j}) + (\mathrm{ad}_r\ (U_j)_+)\tau(2\omega_{j+1,j}).$$

In particular, we can find $c \in U_j$ such that $z'_{j+1} = (\mathrm{ad}_r\ c)\tau(2\omega_{j+1,j})$. Set $z_{j+1} = z'_{j+1}\tau(\omega_{2j+2})$. Since $\tau(\omega_{2j+2})$ is in the center of $\check{U}_j$, the same is true for z_{j+1}.

By (8.19), we see that

$$\tau(2\omega_{j+1,j})\tau(\omega_{2j+2}) = \tau(2\omega_j).$$

Hence

$$\begin{aligned} z_{j+1} =& z'_{j+1}\tau(\omega_{2j+2}) = ((\mathrm{ad}_r\ c)(\tau(2\omega_{j+1,j}))\tau(\omega_{2j+2}) \\ &= (\mathrm{ad}_r\ c)\tau(2\omega_j). \end{aligned}$$

Since $\tau(2\omega_j) \in \check{U}$, and c is in the subalgebra U_j of $\check{U}$, we see that $(\mathrm{ad}_r\ c)\tau(2\omega_j)$ is in $\check{U}$. Thus we have found a central element z'_{j+1} in $\check{U}_j$ and a central element z_{j+1} in $\check{U}_j\check{T}$ such that (i) and (ii) hold.

Recall that we have set $\eta_{t+1} = 0$ and so $\tau(2\eta_{t+1}) = 1$. We can similarly set $\eta'_{j+1} = 0$ and so $\tau(\eta'_{j+1}) = 1$ and $\hat{m}'(2\eta'_{j+1})$ is a nonzero scalar. Suppose first that j is even. Applying Lemma 8.8 (ii) to the quantum symmetric pair $U_j, B \cap U_i$ with $t = j$ and $2l = j$ yields that $\hat{m}'(2\eta'_{j+1})$ is in the $\mathcal{C}$ span of the set (8.20). A similar application using Lemma 8.8 (i) works when j is odd with $t = j$ and $2l - 1 = j$. $\square$

Note that the intersection $[((B\cap U_i)\check{T}_\Theta)_+\check{U}_i\check{T}]\cap\check{U}$ is a subset of $((B\cap U_k)\check{T}_\Theta)_+\check{U}_k\check{T}$ whenever $i \leq k$. Moreover $[((B \cap U_i)\check{T}_\Theta)_+\check{U}_i\check{T}] \cap \check{U}$ is contained in $(B\check{T}_\Theta)_+\check{U}$ for all i. Thus Theorem 8.2 follows once we complete the proof of the next lemma.

LEMMA 8.17. *For each* $1 \leq i \leq t$, *there exists* $b_i \in ((B \cap U_{i-1})\check{T}_\Theta)_+\check{U}_{i-1}\check{T}$ *such that* $\{\tau(2\eta_i) + b_i | 1 \leq i \leq t\}$ *is a subset of* $F_r(\check{U})$.

PROOF. Note that for $j = 1$, we have $((B\cap U_{j-1}\check{T}_\Theta)_+\check{U}_{j-1}\check{T}$ is simply $\mathcal{C}[\check{T}_\Theta]_+\check{U}^0$. By Lemma 2.4, $\tilde{\omega}_1 = \eta_1$. Thus by Lemma 7.8, we can set $b_1 = \tau(2\omega_1) - \tau(2\eta_1)$. We proceed with the proof using induction on j.

Suppose that $1 \leq j < t$. Assume that we have found $b_1, \ldots, b_j$ so that $b_k \in ((B \cap U_{k-1})\check{T}_\Theta)_+\check{U}_{k-1}\check{T}$ for $1 \leq i \leq j$ and $\{\tau(2\eta_k) + b_k | 1 \leq k \leq j\}$ is a subset of $F_r(\check{U})$. By the discussion preceding the lemma, we have $b_k \in ((B \cap U_j)\check{T}_\Theta)_+\check{U}_j\check{T}$ for each k with $1 \leq k \leq j$.

Set $\Sigma'_s = \{\mu_1, \ldots, \mu_j\}$. By Lemma 8.15, for each $1 \leq k \leq j$,

$$\mathcal{P}_B(\mathcal{L}_j(\tau(2\eta_k) + b_k)) = \hat{m}'(2\eta'_k)\tau(k(j+1)^{-1}2\eta_{j+1})$$

up to a nonzero scalar. By the definition of $\mathcal{L}_j$ and Lemma 8.15, $\mathcal{L}_j(\tau(2\eta_k) + b_k)$ is in both $\check{U}_j^{B\cap U_j}$ and $F_r(\check{U})$ for each choice of k. Now $\mathcal{P}_B$ agrees with $\mathcal{P}_{B\cap U_j}$ upon restriction to $\check{U}_j$. Hence Theorem 4.4 implies that the restriction of $\mathcal{P}_B$ to $\check{U}_j^{B\cap U_j}$ is an algebra homomorphism. Therefore

$$\begin{aligned} &\mathcal{P}_B(\mathcal{L}_j(\tau(2\eta_k) + b_k)\mathcal{L}_j(\tau(2\eta_{j+1-k}) + b_{j+1-k})) \\ &\quad = \hat{m}'(2\eta'_k)\hat{m}'(2\eta'_{j+1-k})\tau(2\eta_{j+1}) \end{aligned} \tag{8.21}$$

for all $1 \leq k \leq j$. Moreover, by Lemma 8.15, $\mathcal{L}_j(\tau(2\eta_k) + b_k)\mathcal{L}_j(\tau(2\eta_{j+1-k}) + b_{j+1-k}))$ is in $F_r(\check{U})$ for each k with $1 \leq k \leq j$.

Let z_{j+1} and z'_{j+1} be chosen as in Lemma 8.16. Lemma 8.16(iii) guarantees that 1 is in the $\mathcal{C}$ span of the set

$$\{\mathcal{P}_{B\cap\check{U}_j}(z'_{j+1})\} \cup \{\hat{m}'(2\eta'_k)\hat{m}'(2\eta'_{j+1-k})|\ 1 \leq k \leq j\}.$$

Set $b'_k = b_k\tau(-k(j+1)^{-1}2\eta_{j+1})$ and $b'_{j+1-k} = b_{j+1-k}\tau(-(j+1-k)(j+1)^{-1}2\eta_{j+1})$. Using (8.18), we see that there exists c in $\check{U}_j^{(B\cap U_j)}$ such that c is a linear combination of elements in the set

$$\{z'_{j+1}\} \cup \{\mathcal{L}_j(2\eta'_k + b'_k)\mathcal{L}_j(2\eta'_{j+1-k} + b'_{j+1-k})|\ 1 \leq k \leq j\}$$

and $\mathcal{P}_{(B\cap U_j)}(c) = 1$. By Theorem 5.6, we see that

$$c - 1 \in (B\check{T}_\Theta \cap \check{U}_j)_+\check{U}_j. \tag{8.22}$$

Now consider the element $c\tau(2\eta_{j+1})$. By Lemma 8.16(i) and Lemma 2.4(ii), we have that

$$\mathcal{P}_B(z_{j+1}) = \mathcal{P}_{B\cap\check{U}_j}(z'_{j+1})\tau(2\eta_{j+1}). \tag{8.23}$$

It follows from the choice of c, (8.21), and (8.23), that $c\tau(2\eta_{j+1})$ is a linear combination of elements in the set

$$\{(z_{j+1})\} \cup \{\mathcal{L}_j(2\eta_k + b_k)\mathcal{L}_j(2\eta_{j+1-k} + b_{j+1-k})|\ 1 \le k \le j\}.$$

In particular, $c\tau(2\eta_{j+1}) \in F_r(\check{U})$. Moreover, (8.22) implies that

$$c\tau(2\eta_{j+1}) \in \tau(2\eta_{j+1}) + ((B\cap U_j)\check{T}_\Theta)_+\check{U}_j\check{T}.$$

□

CHAPTER 9

Four Exceptional Cases

In this section, we find the special basis for $\mathcal{P}_B(\check{U}^B)$ described in (7.6) for the four remaining types of symmetric pairs: EIII, EIV, EVII, and EIX. This combined with Theorem 7.9 and Theorem 8.2 completes the proof of the following result (as well as Theorem C of the introduction).

THEOREM 9.1. *Let $\mathfrak{g}, \mathfrak{g}^\theta$ be an irreducible symmetric pair and let $B \in \mathcal{B}$ For each $\gamma \in P^+(\Sigma)$, there exists $b_\gamma \in (B\check{T}_\Theta)_+\check{U}$ such that $\tau(2\gamma) + b_\gamma \in F_r(\check{U})$. Moreover*
$$\{\mathcal{P}_B(\mathcal{L}(\tau(2\gamma) + b_\gamma))|\ \gamma \in P^+(\Sigma)\}$$
is a basis for $\mathcal{P}_B(\check{U}^B)$.

Suppose that B is in $\mathcal{B}$. Recall that $\mathcal{P}_B(Z(\check{U}))$ is a proper subalgebra of $\mathcal{P}_B(\check{U}^B)$ for symmetric pairs of type EIII, EIV, EVII, and EIX (Theorem 6.6). Thus we cannot deduce the dotted W_Θ invariance of $\mathcal{P}_B(\check{U}^B)$ directly from the dotted W_Θ invariance of $\mathcal{P}_B(Z(\check{U}))$ in these cases. Let $c \in \check{U}^B$ and consider the expansion of $\mathcal{P}_B(c)$ as a linear combination of terms of the form $\tau(\beta)$. Using Theorem 5.5 and Lemma 5.8, we see that the the set of terms of the form $\tau(\beta)$ that appear with nonzero coefficient in the $\mathcal{P}_B(c)$ is W_Θ invariant. Using this information combined with the dotted W_Θ invariance of the center, we show that the generators of $\mathcal{P}_B(\check{U}^B)$ are dotted W_Θ invariant. This is the key step in establishing the dotted W_Θ invariance of $\mathcal{P}_B(\check{U}^B)$ for these remaining four problematic types (see Lemma 9.9). The next result, which completes the proof of Theorem A, is an immediate consequence of Lemma 9.9, Theorem 9.1, Theorem 7.7, Theorem 8.1, and Theorem 5.6.

COROLLARY 9.2. *For each irreducible symmetric pair $\mathfrak{g}, \mathfrak{g}^\theta$ and each $B \in \mathcal{B}$, the Harish-Chandra map $\mathcal{P}_B$ maps $\check{U}^B$ onto $\mathcal{C}[\mathcal{A}]^{W_\Theta \circ}$. The kernel of the restriction of $\mathcal{P}_B$ to $\check{U}^B$ is $(B\check{T}_\Theta)_+\check{U} \cap \check{U}^B$. Moreover, $\mathcal{P}_B(Z(\check{U})) = \mathcal{P}_B(\check{U}^B)$ if and only if $\mathfrak{g}, \mathfrak{g}^\theta$ is not of type EIII, EIV, EVII, or EIX.*

The next few lemmas will provide a method for finding appropriate elements as described in (7.10) inside of $F_r(\check{U})$ when $\mathfrak{g}, \mathfrak{g}^\theta$ is of type EIII, EIV, EVII, or EIX. Note that in each of these cases, $\mathfrak{g}$ contains a semisimple Lie subalgebra $\mathfrak{r}$ such that $\mathfrak{r}, \mathfrak{r}^\theta$ is of type DI(1). (In fact, for the latter three cases, $\mathfrak{g}$ contains more than one such Lie subalgebra.) Thus in this subsection, we analyze symmetric pairs of type DI(1). This information is then pulled back to the above four symmetric pair types to finish the proof of Theorem 9.1.

Before turning to the lemmas below, we describe some of the details related to symmetric pairs of type DI(1). In particular, temporarily assume that $\mathfrak{g}, \mathfrak{g}^\theta$ is of type D1(1). Here we have $\mathfrak{g}$ is of type D_n. Assume further that

(9.1) $$\pi_\Theta = \{\alpha_{n-m+1}, \alpha_{n-m+2}, \ldots, \alpha_{n-1}, \alpha_n\}$$

where $n-1 \geq m \geq 3$. The involution Θ on Δ is defined by

$$\begin{aligned} &\Theta(\alpha_i) = \alpha_i \text{ for } \alpha_i \in \pi_\Theta \\ (9.2)\qquad &\Theta(\alpha_i) = -\alpha_i \text{ for } 1 \leq i \leq n-m-1 \\ &\Theta(\alpha_{n-m}) = -(\alpha_{n-m} + 2\alpha_{n-m+1} + \cdots + 2\alpha_{n-2} + \alpha_{n-1} + \alpha_n). \end{aligned}$$

Hence the simple restricted roots are $\tilde{\alpha}_1, \ldots, \tilde{\alpha}_{n-m}$ where

$$\tilde{\alpha}_{n-m} = \alpha_{n-m} + \alpha_{n-m+1} + \cdots + \alpha_{n-2} + (\alpha_{n-1} + \alpha_n)/2 \tag{9.3}$$

$$\tilde{\alpha}_i = \alpha_i \text{ for } 1 \leq i \leq n-m-1. \tag{9.4}$$

Moreover, Σ is a root system of type B_{n-m}.

Recall the definition of the special central elements $z_{2\mu}$ in $\check{U}$ defined immediately after Theorem 6.2. The next lemma gives a detailed description of $\mathcal{P}_B(z_{2\omega_1})$ for symmetric pairs of type D1(1).

LEMMA 9.3. *Let $\mathfrak{g}, \mathfrak{g}^\theta$ be a symmetric pair of type DI(1). Assume further that $\pi_\Theta = \{\alpha_{n-m+1}, \alpha_{n-m+2}, \ldots, \alpha_{n-1}, \alpha_n\}$ where $n-1 \geq m \geq 3$. Then*

$$\mathcal{P}_B(z_{2\omega_1}) = a^{-1}\Big(\sum_{\mu \in W_\Theta \tilde{\omega}_1} q^{(\tilde{\rho}, 2\mu)} \tau(2\mu) + \sum_{i=0}^{m-1} (q^{2i} + q^{-2i}\Big)$$

where

$$a = \Big(\sum_{\mu \in W_\Theta \tilde{\omega}_1} q^{(\tilde{\rho}, 2\mu)} + \sum_{i=0}^{m-1} (q^{2i} + q^{-2i})\Big).$$

PROOF. By [H, Section 13.2, Table 1],

$$\omega_1 = \alpha_1 + \alpha_2 + \cdots + \alpha_{n-2} + (1/2)(\alpha_{n-1} + \alpha_n).$$

Using this table in [H], it is straightforward to see that there does not exist $\beta \in P^+(\pi)$ with $\beta < \omega_1$. Hence it follows from Lemma 6.3 that that

$$\mathcal{P}(z_{2\omega_1}) = a^{-1}\Big(\sum_{\mu \in W\omega_1} q^{(\rho, 2\mu)} \tau(2\mu)\Big)$$

where $a = \sum_{\mu \in W\omega_1} q^{(\rho, 2\mu)}$.

It follows from Lemma 2.1 that $\tilde{\omega}_1 = \omega_1'$, the fundamental weight corresponding to the simple restricted root $\tilde{\alpha}_1$. By [H, Section 13.2], we see that

$$\tilde{\omega}_1 = \omega_1' = \sum_{1 \leq i \leq n-m} \tilde{\alpha}_i. \tag{9.5}$$

From the description of the fundamental weights corresponding to a root system of type B_{n-m}, we further have that the only restricted root in $P^+(\Sigma)$ which is strictly less than ω_1' is 0. Thus the W orbit of ω_1 can be written as a union of two sets

$$S_1 = \{\pm(\alpha_i + \cdots + \alpha_{n-2} + (\alpha_{n-1} + \alpha_n)/2) | 1 \leq i \leq n-m\}$$

and

$$S_2 = \{\pm(\alpha_i + \alpha_{i+1} + \cdots + \alpha_n - (\alpha_{n-1} + \alpha_n)/2) |\ n-m+1 \leq i \leq n-1\}$$

It is straightforward to check that $\mu = \tilde{\mu}$ and hence $(\rho, \mu) = (\tilde{\rho}, \mu)$ for all $\mu \in S_1$. Moreover, the set of $\tilde{\mu}$ with $\mu \in S_1$ corresponds to the W_Θ orbit of ω_1'. On the other hand, S_2 is a subset of $Q(\pi_\Theta)$. Hence, the image under $\tilde{}$ of elements in the second set S_2 are all equal to 0.

We have

$$\mathcal{P}_B(z_{2\omega_1}) = a^{-1}(\sum_{\mu\in S_1} q^{(\tilde{\rho},2\mu)}\tau(2\mu) + \sum_{\mu\in S_2} q^{(\rho,2\mu)}).$$

The lemma now follows from the fact that

$$\sum_{\mu\in S_2} q^{(\rho,2\mu)} = \sum_{i=0}^{m-1}(q^{2i}+q^{-2i}).$$

□

We continue to study the symmetric pair $\mathfrak{g}, \mathfrak{g}^\theta$ of type DI(1) with π_Θ given by (9.1). Consider the fundamental weight ω_n in $P^+(\pi)$. By Section 2, Case 2 and the formula for $\Theta(\alpha_{n-m})$ given in (9.2), we have that $\tilde{\omega}_n = \omega'_{n-m}$. Set

$$\eta = \omega'_{n-m}.$$

By [H, Section 13.2, Table 1], we have

$$\tilde{\omega}_n = \eta = 1/2(\tilde{\alpha}_1 + 2\tilde{\alpha}_2 + \cdots + (n-m)\tilde{\alpha}_{n-m}). \tag{9.6}$$

Recall that $\varphi_{2\eta}$ is the image of the zonal spherical function $g_{2\eta}$ inside the W_Θ ring $\mathcal{C}[P(2\Sigma)]^{W_\Theta}$ (see Section 1). Set $\varphi = \varphi_{2\eta}$. Multiplying by a nonzero scalar if necessary, we may assume that $\varphi(\tau(\tilde{\rho})) = 1$. Using the table of fundamental weights in [H, Section 13.2] for root systems of type B_{n-m}, one checks that there are no restricted weights in $P^+(\Sigma)$ strictly less than η. Hence

$$\varphi = \frac{\sum_{\mu\in W_\Theta\eta} z^{2\mu}}{\sum_{\mu\in W_\Theta\eta} q^{(\tilde{\rho},2\mu)}}. \tag{9.7}$$

Set $c_{2\omega_1} = \mathcal{L}(\tau(2\omega_1))$. In the next lemma, we use φ to distinguish between the B invariant element $c_{2\omega_1}$ and the central element $z_{2\omega_1}$.

LEMMA 9.4. *Suppose that $\mathfrak{g}, \mathfrak{g}^\theta$ is of type DI(1). Assume further that $\pi_\Theta = \{\alpha_{n-m+1}, \alpha_{m-m+2}, \ldots, \alpha_n\}$ with $n-1 \geq m \geq 3$. Then $\mathcal{P}_B(c_{2\omega_1}) \neq \mathcal{P}_B(z_{2\omega_1})$.*

PROOF. By Lemma 7.5 we have

$$\varphi(\tau(2\omega_1)\tau(\tilde{\rho})) = z^{2\eta}(\mathcal{P}_B(c_{2\omega_1})).$$

One checks using the formula for η given in (9.6) that $\sum_{\mu\in W_\Theta\eta} z^{2\mu}$ is equal to the product

$$\prod_{m\leq s\leq n-1} (z^{(\tilde{\alpha}_{n-s}+\tilde{\alpha}_{n-s+1}+\cdots+\tilde{\alpha}_{n-m})} - z^{-(\tilde{\alpha}_{n-s}+\tilde{\alpha}_{n-s+1}+\cdots+\tilde{\alpha}_{n-m})}). \tag{9.8}$$

Indeed, writing (9.8) as a sum of elements of the form z^β, we see that (9.8) is an element of the set

$$z^{2\eta} + \sum_{\gamma\in Q^+(\Sigma)} \mathbf{N} z^{2\eta-2\gamma}.$$

On the other hand, it is not hard to see that (9.8) is W_Θ invariant. Since there are no weights in $P^+(\Sigma)$ strictly less than η, the desired equality follows.

By (9.3) and (9.4) we have $(\tilde{\rho}, \tilde{\alpha}_{n-m}) = (\rho, \tilde{\alpha}_{n-m}) = m$ while $(\tilde{\rho}, \tilde{\alpha}_i) = (\rho, \alpha_i) = 1$ for $1 \leq i \leq n-m-1$. It follows from (9.7) and the previous paragraph that

$$\varphi = a^{-1} \prod_{m\leq s\leq n-1} (z^{(\tilde{\alpha}_{n-s}+\tilde{\alpha}_{n-s+1}+\cdots+\tilde{\alpha}_{n-m})} - z^{-(\tilde{\alpha}_{n-s}+\tilde{\alpha}_{n-s+1}+\cdots+\tilde{\alpha}_{n-m})})$$

where

$$a = \prod_{m \le s \le n-1} (q^s + q^{-s}).$$

Now (9.3) and (9.4) and the fact that π generates a root system of type D_n ensure that $(\omega_1, \tilde{\alpha}_i) = \delta_{i1}(\omega_1, \alpha_1) = \delta_{ij}$. Hence

$$\begin{aligned}\varphi(\tau(2\omega_1 + \tilde{\rho})) &= a^{-1}(q^{n+1} + q^{-n-1}) \prod_{m \le s \le n-2} (q^s + q^{-s}) \\ &= (q^{n+1} + q^{-n-1})(q^{n-1} - q^{-n+1})^{-1}.\end{aligned}$$

Furthermore, it is straightforward to check using (9.5) that the W_Θ orbit of $\omega_1' = \tilde{\omega}_1$ is the set $\{\pm(\tilde{\alpha}_{n-s} + \tilde{\alpha}_{n-s+1} + \cdots + \tilde{\alpha}_{n-m}) | \; m \le s \le n-1\}$. By Lemma 9.3, $b(\mathcal{P}_B(z_{2\omega_1}) - \sum_{i=0}^{m-1}(q^{2i} + q^{-2i}))$ equals

$$\sum_{m \le s \le n-1} q^{2s}\tau(2(\tilde{\alpha}_{n-s} + \cdots + \tilde{\alpha}_{n-m})) + q^{-2s}\tau(-2(\tilde{\alpha}_i + \cdots + \tilde{\alpha}_{n-m}))$$

where $b = (\sum_{m \le s \le n-1} q^{2s} + q^{-2s}) + \sum_{i=0}^{m-1}(q^{2i} + q^{-2i})$. Hence using (9.6) we have

$$z^{2\eta}(\mathcal{P}_B(z_{2\omega_1})) = b^{-1}(q^{2n} + q^{-2n} - (q^{2(n-1)} + q^{-2(n-1)}) + b).$$

Thus $\mathcal{P}_B(z_{2\omega_1}) = \mathcal{P}_B(c_{2\omega_1})$ implies that

$$(q^{n+1} + q^{-n-1})b = (q^{2n} + q^{-2n} - (q^{2(n-1)} + q^{-2(n-1)}) + b)(q^{n-1} + q^{-n+1}). \tag{9.9}$$

Note that the coefficient of q^{n-1} is 0 in the right hand side of (9.9). On the other hand, the coefficient of q^{n-1} in the left hand side of (9.9) is 2. This contradiction proves the lemma. □

We now consider the three exceptional types of symmetric pairs, EIV, EVII, and EIX. Recall the description of the involution Θ provided before Lemma 2.5 and the list of simple restricted roots given in (2.9). By Lemma 2.5(ii), we have that the set of fundamental weights in $P^+(\Sigma)$ not contained in the image of $P^+(\pi)$ under $\tilde{}$ is precisely the set $\{\omega_1', \omega_6'\}$. Furthermore, one checks that ω_6' is the unique nonzero element of $P^+(\Sigma)$ strictly less than $\tilde{\omega}_1 = 2\omega_1'$ while ω_1' is the unique nonzero element of $P^+(\Sigma)$ strictly less than $\tilde{\omega}_n$. Here $\tilde{\omega}_n = \omega_n'$ in the latter two cases (i.e. $n = 7$ or $n = 8$) while $\tilde{\omega}_n = 2\omega_6'$ when $\mathfrak{g}, \mathfrak{g}^\theta$ is of type EIV.

Theorem 9.1 follows from Theorem 7.7, Lemma 7.8, and the next lemma for symmetric pairs of type EIV, EVII, and EIX.

LEMMA 9.5. *Assume that $\mathfrak{g}, \mathfrak{g}^\theta$ is of type EIV, EVII, or EIX. Then there exists $f \in (\mathrm{ad}_r\ U)\tau(2\omega_1)$ such that*

$$f \in \tau(2\omega_6') + (B\check{T}_\Theta)_+\check{U}$$

and $g \in (\mathrm{ad}_r\ U)\tau(2\omega_n)$ such that

$$g \in \tau(2\omega_1') + (B\check{T}_\Theta)_+\check{U}.$$

PROOF. Let π' be the subset of π equal to $\{\alpha_i | \; i > 1\}$. Note that π' generates a root system of type D_{n-1}. Furthermore, α_{n-i} is the $(i+1)^{th}$ simple root in this root system with respect to the ordering of the simple roots given in [H, Chapter III]. Now Θ restricts to an involution on the root system generated by π'. Moreover, $\pi' \cap \pi_\Theta = \pi_\Theta = \{\alpha_2, \alpha_3, \alpha_4, \alpha_5\}$. Let $\mathfrak{r}$ be the semisimple Lie subalgebra of $\mathfrak{g}$ generated by the positive and negative root vectors corresponding to the simple roots in π'. The Lie algebra $\mathfrak{r}$ has rank $n-1$ and $\mathfrak{r}, \mathfrak{r}^\theta$ is a symmetric pair of type

DI(1). (This is easy to see by directly comparing the description of Θ for symmetric pairs of type EIV,EVII, and EIX given right before Lemma 2.5 to the description of Θ involution for symmetric pairs of type D1(1) stated in (9.2).) We write $U_q(\mathfrak{r})$ for the quantized enveloping algebra of $\mathfrak{r}$ identified in the obvious way with a subalgebra of U. Let $\check{U}_q(\mathfrak{r})$ denote the simply connected quantized enveloping algebra of $\mathfrak{r}$. Let Σ' be the restricted root system associated to the symmetric pair $\mathfrak{r}, \mathfrak{r}^\theta$ and let W'_Θ denote the corresponding restricted Weyl group.

Let ν_n denote the fundamental weight associated to the simple root α_n considered as an element in $\mathbf{Q}\Sigma'$, the $\mathbf{Q}$ vector space generated by the simple roots in π'. Let $z'_{2\nu_n}$ be the unique central element of $\check{U}_q(\mathfrak{r})$ such that

$$z'_{2\nu_n} \in \tau(2\nu_n) + (\mathrm{ad}_r\ U_q(\mathfrak{r})_+)\tau(2\nu_n). \tag{9.10}$$

Let $\mathcal{L}'$ be the map defined by Definition 7.3 and (7.5) for the quantum symmetric pair $U_q(\mathfrak{r}), B \cap U_q(\mathfrak{r})$. Set $c'_{2\nu_n} = \mathcal{L}'(\tau(2\nu_n))$. Recall that $\kappa(t) = t$ for all $t \in \check{T}$ (see Section 1 and (1.22)). By Definition 7.3 and (7.5) we further have that $c'_{2\nu_n}$ is the unique $(U_q(\mathfrak{r}) \cap B)$ invariant element of $\check{U}_q(\mathfrak{r})$ such that

$$c'_{2\nu_n} \in \tau(2\nu_n) + \kappa((\mathrm{ad}_r\ (U_q(\mathfrak{r}) \cap B)_+)\tau(2\nu_n)). \tag{9.11}$$

It follows from the relationship between the right adjoint action and κ expressed in (1.24) that

$$\kappa((\mathrm{ad}_r\ a)b) \in (\mathrm{ad}_r\ U_q(\mathfrak{r}))\kappa(b)$$

for all $a \in U_q(\mathfrak{r})$ and $b \in \check{U}$. In particular, $c'_{2\nu_n}$ is an element of the space $(\mathrm{ad}_r\ U_q(\mathfrak{r}))(\tau(2\nu_n))$. Note that $z'_{2\nu_n}$ is also an element of $(\mathrm{ad}_r\ U_q(\mathfrak{r}))\tau(2\nu_n)$.

Now Σ' is a root system of type B_{n-5}. Moreover, if we order the roots of Σ' as in [H], then $\tilde{\alpha}_n$ corresponds to the first simple root of Σ'. By (9.5), $\tilde{\nu}_n$ is the fundamental weight corresponding to $\tilde{\alpha}_n$. It follows that the only weight in $P^+(\Sigma')$ less than $\tilde{\nu}_n$ is 0. This fact combined with (6.2) and subsequent discussion implies that $\mathcal{P}_{B\cap U_q(\mathfrak{r})}(z'_{2\nu_n})$ is a linear combination of 1 and $\sum_{\gamma \in W'_\Theta \nu_n} q^{(\rho,2\gamma)}\tau(2\gamma)$. Moreover, the coefficient of $\sum_{\gamma \in W'_\Theta \nu_n} q^{(\rho,2\gamma)}\tau(2\gamma)$. is nonzero. A similar statement can be made concerning $\mathcal{P}_{B\cap U_q(\mathfrak{r})}(c'_{2\eta_n})$. In particular, since $\mathfrak{r}, \mathfrak{r}^\theta$ is a symmetric pair of type DI(1), we may apply Theorem 6.5 to see that $\mathcal{P}_{B\cap U_q(\mathfrak{r})}(c'_{2\eta_n})$ is invariant under the action of W_Θ. Lemma 7.6 and the fact that 0 is the only weight in $P^+(\Sigma')$ less than $\tilde{\nu}_n$ thus imply that $\mathcal{P}_{B\cap U_q(\mathfrak{r})}(c'_{2\eta_n})$ is also a linear combination of 1 and $\sum_{\gamma \in W'_\Theta \nu_n} q^{(\rho,2\gamma)}\tau(2\gamma)$. By Lemma 7.5, we further have that $\mathcal{P}_{B\cap U_q(\mathfrak{r})}(c'_{2\eta_n})$ is nonzero.

Suppose that

$$\mathcal{P}_{B\cap U_q(\mathfrak{r})}(z'_{2\nu_n}) = d\mathcal{P}_{B\cap U_q(\mathfrak{r})}(c'_{2\nu_n})$$

for some nonzero scalar d. Lemma 9.4 ensures that

$$\mathcal{P}_{B\cap U_q(\mathfrak{r})}(z'_{2\nu_n}) \neq \mathcal{P}_{B\cap U_q(\mathfrak{r})}(c'_{2\nu_n}).$$

Hence $d \neq 1$. By Theorem 5.6, it follows that

$$dz'_{2\nu_n} - c'_{2\nu_n} \in (B\check{T}_\Theta)_+\check{U}.$$

In particular, $dz'_{2\nu_n} - c'_{2\nu_n}$ is an element of $\check{U}_+$. On the other hand, by (9.10) and (9.11) we have that $z'_{2\nu_n}$ is not in $\check{U}_+$ and $z'_{2\nu_n} - c'_{2\nu_n} \in \check{U}_+$. This forces $d = 1$, a contradiction. Thus $\mathcal{P}_{B\cap U_q(\mathfrak{r})}(z'_{2\nu_n})$ is a not a scalar multiple of $\mathcal{P}_{B\cap U_q(\mathfrak{r})}(c'_{2\nu_n})$. It follows that there is a linear combination X of $z'_{2\nu_n}$ and $c'_{2\nu_n}$ such that $\mathcal{P}_{B\cap U_q(\mathfrak{r})}(X) =$

1. Hence $\mathcal{P}_{B\cap U_q(\mathfrak{r})}(X-1)=0$. Thus by Theorem 5.6, $X-1\in(B\check{T}_\Theta\cap\check{U}_q(\mathfrak{r}))_+U$. In particular,

$$X\in(\mathrm{ad}_r\ U_q(\mathfrak{r}))\tau(2\nu_n)\text{ and }X\in 1+(B\check{T}_\Theta)_+\check{U}. \tag{9.12}$$

Since α_n is the first root in the root system of type D_{n-2} generated by π', we have

$$\nu_n=1/2(\alpha_2+\alpha_3)+\alpha_4+\cdots+\alpha_n.$$

Note that $(\omega_n-\nu_n,\alpha_i)=0$ for all $\alpha_i\in\pi'$. On the other hand

$$(\omega_n-\nu_n,\alpha_1)=(-\nu_n,\alpha_1)=-(\alpha_3,\alpha_1)/2=(\omega_1,\alpha_1)/2.$$

It follows that $\omega_n=\nu_n+\omega_1/2$. Now $\tau(\omega_1)$ commutes with elements of $U_q(\mathfrak{r})$. Hence

$$(\mathrm{ad}_r\ U_q(\mathfrak{r}))\tau(2\omega_n)=((\mathrm{ad}_r\ U_q(\mathfrak{r}))\tau(2\nu_n))\tau(\omega_1). \tag{9.13}$$

It follows from (9.12) and (9.13) that

$$X\tau(\omega_1)\in(\mathrm{ad}_r\ U)\tau(2\omega_n)\text{ and }X\tau(\omega_1)\in\tau(\tilde{\omega}_1)+(B\check{T}_\Theta)_+\check{U}.$$

The second assertion of the lemma now follows from the fact that $\tilde{\omega}_1=2\omega_1'$ (Lemma 2.5(ii)). The first assertion is proved in exactly the same way where we replace π' with the set $\{\alpha_i|\ i<5\}$. □

We next turn our attention to the symmetric pair of type EIII. Recall the list of restricted simple roots (2.8). Note further that both $\tilde{\alpha}_1$ and $2\tilde{\alpha}_1$ are elements of Σ. In particular, the restricted root system Σ is nonreduced of type BC_2. Now Σ contains a root system with set of positive simple roots $\{\tilde{\alpha}_2,2\tilde{\alpha}_1\}$ of type C_2. Here $\tilde{\alpha}_2$ is the short simple root and $2\tilde{\alpha}_1$ is the long simple root. As explained at the beginning of Section 2, the fundamental weight ω_1' associated to $\tilde{\alpha}_1$ satisfies $(\omega_1',2\tilde{\alpha}_1)=(2\tilde{\alpha}_1,2\tilde{\alpha}_1)/2$. In particular, the weight lattice $P(\Sigma)$ is the same as the weight lattice of the underlying root system of type C_2. Thus $\omega_1'=2\tilde{\alpha}_1+\tilde{\alpha}_2$ and $\omega_2'=\tilde{\alpha}_1+\tilde{\alpha}_2$. By Lemma 2.5(i), we have $\tilde{\omega}_1=\omega_1'$ and $\tilde{\omega}_2=2\omega_2'$. In particular, recall that $\omega_1'\in\widetilde{P^+(\pi)}$ while ω_2' is not an element of this set.

By Lemma 7.8, we have that $\tau(2\omega_1')+(\tau(2\omega_1)-\tau(2\omega_1'))\in F_r(\check{U})$ and $\tau(2\omega_1)-\tau(2\omega_1')\in(B\check{T}_\Theta)_+\check{U}$. Thus Theorem 9.1 for this last case follows from the next lemma.

LEMMA 9.6. *Assume that $\mathfrak{g},\mathfrak{g}^\theta$ is of type EIII. Then there exists $f\in F_r(\check{U})$ such that*

$$f\in\tau(2\omega_2')+(B\check{T}_\Theta)_+\check{U}.$$

PROOF. Let π' be the subset of π equal to $\{\alpha_2,\alpha_3,\alpha_4,\alpha_5\}$, Note that π' generates a root system of type D_4. Furthermore, α_i is the $(i-1)^{th}$ simple root with respect to the ordering of the simple roots given in [H]. Now Θ restricts to an involution on the root system generated by π'. Note that $\pi'\cap\pi_\Theta=\{\alpha_3,\alpha_4,\alpha_5\}$. Let ν_2 denote the fundamental weight associated to the simple root α_2 contained in the rational span of the roots in π'. Using the table in [H, Section 13.2] for a root system of type D_4, we see that $\nu_2=(\alpha_3+\alpha_5)/2+\alpha_4+\alpha_2$. Now $(\omega_2-\nu_2,\alpha_i)=(\omega_i,\alpha_i)/2$ for $i=1$ and $i=6$. Also, $(\omega_2-\nu_2,\alpha_i)=0$ for $i\notin\{1,6\}$. Therefore

$$\omega_2=\nu_2+(\omega_1+\omega_6)/2.$$

By Lemma 2.5(i), $\tilde{\omega}_1+\tilde{\omega}_6=2\omega_1'$. The rest of the argument follows as in the proof of Lemma 9.5. □

In order to complete the proof of Corollary 9.2, we must show that $\mathcal{P}_B(\check{U}^B)$ is invariant under the dotted action of W_Θ. Suppose that $X = \sum_\gamma a_\gamma \tau(\gamma)$ is an element of $\mathcal{C}(Q(\Sigma))[\mathcal{A}]$ where the a_γ are in $\mathcal{C}(Q(\Sigma))$. Define the support of X, denoted by $\mathrm{Supp}(X)$, as the following set

$$\mathrm{Supp}(X) = \{\tau(\gamma) |\ a_\gamma \neq 0\}.$$

The next step in establishing the desired dotted W_Θ invariance is to show that the support of elements in $\check{U}^B$ is W_Θ invariant. In particular, we have the following result.

LEMMA 9.7. *The support of $\mathcal{P}_B(c)$ is W_Θ invariant for each $c \in \check{U}^B$.*

PROOF. Let $c \in \check{U}^B$. By Lemma 5.8,

$$\mathrm{Supp}(\mathcal{P}_B(c)) = \mathrm{Supp}(\mathcal{X}(c))$$

for all $c \in \check{U}^B$. By Theorem 5.5, we have that $\mathcal{X}(c)$ is W_Θ invariant. Hence $\mathrm{Supp}(\mathcal{P}_B(c))$ is W_Θ invariant. □

By Theorem 9.1, we can find a set of elements $\{b_\gamma |\ \gamma \in P^+(\Sigma)\}$ in $(B\check{T}_\Theta)_+\check{U}$ such that $\tau(2\gamma) + b_\gamma \in F_r(\check{U})$. Set

$$c_{2\eta} = \mathcal{L}(\tau(2\eta) + b_\eta)$$

for each $\eta \in P^+(\Sigma)$. Recall the degree function defined on $\mathcal{C}[\mathcal{A}]$ in Section 5 (see (5.18)) and the notion of top(X) as the top homogeneous term of an element $X \in \mathcal{C}[\mathcal{A}]$. By the proof of Theorem 7.7 (see (7.9), we have that

$$\mathrm{top}(c_{2\eta}) = \tau(2\eta) \tag{9.14}$$

up to a nonzero scalar for each choice of η. Recall the definition of $\hat{m}(2\eta)$ given in (6.5).

LEMMA 9.8. *Let $\eta \in P^+(\Sigma)$ and assume that there exists λ and μ in $P^+(\pi)$ such that $\tilde{\lambda} = \tilde{\mu} + \eta$. Then there exists a nonzero scalar a such that*

$$\mathcal{P}_B(c_{2\eta}) \in a\hat{m}(2\eta) + \sum_{\{\gamma \in P^+(\Sigma) | \gamma < \eta\}} \sum_{w \in W_\Theta} \mathcal{C}\tau(2w\gamma).$$

Moreover, if the only weights in $P^+(\Sigma)$ less than η are contained in the set $\{0\}$, then $\mathcal{P}_B(c_{2\eta})$ is invariant under the dotted W_Θ action.

PROOF. Recall Theorem 4.6 that $\mathcal{P}_B(\check{U}^B)$ is a subalgebra of $\mathcal{C}[\mathcal{A}]$. By Lemma 9.7 and (9.14), we can find nonzero scalars $a_{2\beta}$ that

$$\mathcal{P}_B(c_{2\eta}) \in \sum_{\beta \in W_\Theta \eta} a_{2\beta} q^{(\rho, 2\beta)} \tau(2\beta) + \sum_{\{\gamma \in P^+(\Sigma) | \gamma < \eta\}} \sum_{w \in W_\Theta} \mathcal{C}\tau(2w\gamma). \tag{9.15}$$

Now consider the central elements $z_{2\lambda}$ and $z_{2\mu}$. By the discussion following Lemma 6.3 (see in particular (6.3)) we have that

$$\mathrm{top}(\mathcal{P}_B(z_{2\mu})) = eq^{(\rho, 2\tilde{\mu})}\tau(2\tilde{\mu})$$

for some nonzero scalar e. The dotted W_Θ invariance of the image of the center under $\mathcal{P}_B$ (Theorem 6.2) implies that

$$\mathcal{P}_B(z_{2\mu}) \in e\hat{m}(2\tilde{\mu}) + \sum_{\{\gamma \in P^+(\Sigma) | \gamma < \tilde{\mu}\}} \sum_{w \in W_\Theta} \mathcal{C}\tau(2w\gamma).$$

Similarly,

$$\mathcal{P}_B(z_{2\lambda}) \in f\hat{m}(2\tilde{\lambda}) + \sum_{\{\gamma\in P^+(\Sigma)|\gamma<\tilde{\lambda}\}} \sum_{w\in W_\Theta} \mathcal{C}\tau(2w\gamma)$$

for some nonzero scalar f.

Consider the element $ea_{2\eta}z_{2\lambda} - fc_{2\eta}z_{2\mu}$ in $\check{U}^B$. We can find scalars d_β such that

$$\mathcal{P}_B(ea_{2\eta}z_{2\lambda} - fc_{2\eta}z_{2\mu}) \in \sum_{\beta\in W_\Theta\tilde{\lambda}} d_\beta\tau(2\beta) + \sum_{\{\gamma\in P^+(\Sigma)|\gamma<\tilde{\lambda}\}} \sum_{w\in W_\Theta} \mathcal{C}\tau(2w\gamma).$$

Note that the coefficient $d_{\tilde{\lambda}}$ of $\tau(2\tilde{\lambda})$ is zero. It follows from Lemma 9.7 that $d_\beta = 0$ all $\beta \in W_\Theta\tilde{\lambda}$. This forces $a_{2\beta} = a_{2\eta}$ for each $\beta \in W_\Theta\eta$. The main assertion of the lemma now follows by setting $a = a_{2\eta}$.

Now assume in addition that the only weights less than η are contained in the set $\{0\}$. This assumption and the above argument permits us to rewrite (9.15) as

$$\mathcal{P}_B(c_{2\eta}) \in a_{2\eta}\hat{m}(2\eta) + \mathcal{C}.$$

In particular $\mathcal{P}_B(c_{2\eta})$ is invariant under the dotted W_Θ action. □

Since $\mathcal{P}_B$ is an algebra homomorphism, we have that

$$\text{top}(\mathcal{P}_B(ab)) = \text{top}(\mathcal{P}_B(a))\text{top}(\mathcal{P}_B(b)) \tag{9.16}$$

for all a and b in $\check{U}^B$. Set t equal to the rank of Σ and let $\eta_1, \ldots, \eta_t$ denote the fundamental restricted weights. Assume for the moment that we have found elements $c_1, \ldots, c_t$ such that $\text{top}(\mathcal{P}_B(c_i)) = \tau(2\eta_i)$ and $\mathcal{P}_B(c_i)$ is invariant under the dotted W_Θ action for each i . By (9.16) and Theorem 5.9, the elements $\mathcal{P}_B(c_1), \ldots, \mathcal{P}_B(c_t)$ generate $\mathcal{P}_B(\check{U}^B)$, and so $\mathcal{P}_B(\check{U}^B)$ is W_{Θ° invariant. We find such elements c_i, $1 \le i \le t$, for the four symmetric pairs of this section in the next lemma. This completes the proof of Corollary 9.2.

LEMMA 9.9. *For each i such that $1 \le i \le t$, there exists $c_i \in \check{U}^B$ such that*

(i) $\text{top}(\mathcal{P}_B(c_i)) = \tau(2\eta_i)$ *up to a nonzero scalar.*
(ii) $\mathcal{P}_B(c_i) \in \mathcal{C}[\mathcal{A}]^{W_{\Theta^\circ}}$.

PROOF. Suppose first that the fundamental restricted weight η_i satisfies $\eta_i = \tilde{\omega}$ where $\omega \in P^+(\pi)$. Set $c_i = z_{2\omega}$. By the discussion following Lemma 6.3 and (6.3), we have that c_i satisfies (i) while (ii) follows from Theorem 6.2. Now suppose that η_i is not contained in $\widetilde{P^+(\pi)}$. Set $c_i = c_{2\eta_i}$. Assume there exists λ and μ in $P^+(\pi)$ such that $\eta_i + \tilde{\mu} = \tilde{\lambda}$. Assume further that the only elements in $P^+(\Sigma)$ less than η_i are contained in the set $\{0\}$. Then by (9.14) and Lemma 9.8, $c_{2\eta_i}$ satisfies conditions (i) and (ii). These two cases are enough to establish the lemma for symmetric pairs of type EIII, EIV, and EVII. A little more work is needed in type EIX. Here are the details.

Assume first that $\mathfrak{g}, \mathfrak{g}^\theta$ is of type EIII. Recall the discussion preceding Lemma 9.6. In particular, $t = 2$ and the fundamental weights are $\eta_1 = \omega_1' = 2\tilde{\alpha}_1 + \tilde{\alpha}_2$ and $\eta_2 = \omega_2' = \tilde{\alpha}_1 + \tilde{\alpha}_2$. Since $\eta_1 = \tilde{\omega}_1$ we may set $c_1 = z_{2\omega_1}$. Note that η_1 is not less than η_2 and so the only element in $P^+(\Sigma)$ less than η_2 is 0. By Lemma 2.5(i), we see that $\eta_2 + \tilde{\mu} = \tilde{\lambda}$ where $\mu = \omega_1$ and $\lambda = \omega_3$. Hence we may set $c_2 = c_{2\eta_2}$.

Now assume that $\mathfrak{g}, \mathfrak{g}^\theta$ is of type EIV, EVII, or EIX. By Lemma 2.5(ii), the fundamental weights in $P^+(\Sigma)$ are $\eta_1 = \omega_1'$, $\eta_2 = \omega_6'$, and $\eta_i = \omega_{i+4}'$ for $i \ge 3$.

By Lemma 2.5, η_i is in $\widetilde{P^+(\pi)}$ for $i \geq 3$. Hence we may set $c_i = z_{2\eta_i}$ for $i \geq 3$. Moreover, using Lemma 2.5 (ii) we see that $\omega_1' + \tilde{\omega}_6 = \tilde{\omega}_5$ and $\omega_6' + \tilde{\omega}_1 = \tilde{\omega}_3$. We argue below for each of these three cases that we may choose $c_1 = c_{2\eta_1}$ and $c_2 = c_{2\eta_2}$. The root system type of Σ can be found in the list in [A] or in [L5, Appendix A].

$\mathfrak{g}, \mathfrak{g}^\theta$ is of type EIV: Here, Σ is a root system of type A_2. Hence both η_1 and η_2 are minimal weights in $P^+(\Sigma)$ with respect to the standard partial ordering. Thus by Lemma 9.8, we may set $c_1 = c_{2\eta_1}$ and $c_2 = c_{2\eta_2}$.

$\mathfrak{g}, \mathfrak{g}^\theta$ is of type EVII: The restricted root system Σ is of type C_3. Now the ordering of the fundamental restricted weights given here correspond to the ordering for a root system of type C_3 given in [H]. Hence by [H, Section 13.2], the only weights in $P^+(\Sigma)$ less than η_i for $i = 1, 2$ are contained in the set $\{0\}$. Thus, by Lemma 9.8, we may set $c_1 = c_{2\eta_1}$ and $c_2 = c_{2\eta_2}$.

$\mathfrak{g}, \mathfrak{g}^\theta$ is of type EIX: The restricted root system is of type F_4. Here, the ordering $\eta_1, \eta_2, \eta_3, \eta_4$ of the fundamental weights described above is in the reverse ordering of the fundamental weights for the descriptionof F_4 given in [H]. In particular, checking the table of fundamental weights in [H, Section 13.2], we see that the only weight in $P^+(\Sigma)$ less than η_1 is 0. Hence by Lemma 9.8, $c_{2\eta_1}$ satisfies conditions (i) and (ii) of the lemma. It follows that

$$\mathcal{P}_B(c_{2\eta_1}) \in d\hat{m}(2\eta_1) + \mathcal{C}$$

where d is a nonzero scalar. The situation for η_2 is more complicated. One checks that the only dominant integral weights less than η_2 are η_1 and η_4. Moreover, $\eta_1 < \eta_4$. By Lemma 9.8, we can write

$$\mathcal{P}_B(c_{2\eta_2}) \in a\hat{m}(2\eta_2) + \sum_{\gamma \in W_\Theta \eta_4} \mathcal{C}q^{(\rho, 2\gamma)}\tau(2\gamma) + \sum_{\gamma \in W_\Theta \eta_1} \mathcal{C}q^{(\rho, 2\gamma)}\tau(2\gamma) + \mathcal{C}$$

where a is a nonzero scalar. Now $\eta_4 = \tilde{\omega}_8$ and the weights in $P^+(\Sigma)$ less than η_4 are precisely 0 and η_1. So

$$\mathcal{P}_B(z_{2\omega_8}) \in b\hat{m}(2\eta_4) + \mathcal{C}\hat{m}(2\eta_1) + \mathcal{C}$$

where b is a nonzero scalar. It follows that there exists a linear combination X of $c_{2\eta_1}$, and $z_{2\omega_8}$ such that

$$\begin{aligned} \mathcal{P}_B(c_{2\eta_2} - X) \in a\hat{m}(2\eta_2) &+ \sum_{\gamma \in W_\Theta \eta_4} \mathcal{C}q^{(\rho, 2\gamma)}\tau(2\gamma) \\ &+ \sum_{\gamma \in W_\Theta \eta_1} \mathcal{C}q^{(\rho, 2\gamma)}\tau(2\gamma) + \mathcal{C} \end{aligned} \tag{9.17}$$

with the coefficient of $\tau(2\eta_1)$ and the coefficient of $\tau(2\eta_4)$ in (9.17) is equal to zero. By Lemma 9.7, the support of $\mathcal{P}_B(c_{2\eta_2} - X)$ is W_Θ invariant. Hence the coefficients of $\tau(2w\eta_1)$ and $\tau(2w\eta_4)$ in (9.17) are all equal to zero as w runs over elements of W_Θ. Thus

$$\mathcal{P}_B(c_{2\eta_2} - X) \in a\hat{m}(2\eta_2) + \mathcal{C}.$$

It follows that $\mathcal{P}_B(c_{2\eta_2} - X)$ is invariant under the dotted W_Θ action. By the choice of X, we also have that $\mathcal{P}_B(X)$ is dotted W_Θ invariant. Hence $\mathcal{P}_B(c_{2\eta_2})$ is invariant under the dotted W_Θ action. Moreover, the top degree term of $\mathcal{P}_B(c_{2\eta_2})$ equals $aq^{(\rho, 2\eta_2)}\tau(2\eta_2)$. Thus $c_2 = c_{2\eta_2}$ satisfies (i) and (ii). $\square$

Appendix: Commonly Used Notation

Here is a list of notation defined in Section 1 (in the following order): $\mathbf{C}$, $\mathbf{Q}$, $\mathbf{Z}$, $\mathbf{R}$, $\mathbf{N}$, q, $\mathcal{C}$, $\mathcal{R}$, $Q(\Phi)$, $Q^+(\Phi)$, $P(\Phi)$ $P^+(\Phi)$, $\mathfrak{g}$, $\mathfrak{n}^- \oplus \mathfrak{h} \oplus \mathfrak{n}^+$, Δ, $(\ ,\ )$, $\pi = \{\alpha_1, \ldots, \alpha_n\}$, $\leq$, e_i, f_i, h_i, θ, $\mathfrak{g}^\theta$ Σ, Θ, $\tilde{\alpha}$, π_Θ p(i), π^*, q_i, $U_q(\mathfrak{g})$, U, x_i, y_i, $t_i^{\pm 1}$, T, U^+, G^-, U^0, $\mathit{\Delta}$, σ, ϵ, $(\mathrm{ad}_r\ a)b$, $(\mathrm{ad}\ a)b$, τ, S_β, A_+, $\check{U}$, $\check{T}$, $\check{U}^0$, $F_r(\check{U})$, z^λ, $L(\lambda)$, $\mathcal{C}[G]$ (G a multiplicative group), $\mathcal{C}[H]$ (H an additive group), B_θ, $\mathcal{M}$, T_Θ, $\tilde{\theta}$, B_i, $\mathcal{S}$, $\mathcal{D}$, $U_\mathcal{R}$, $\mathbf{H}$, $\mathcal{B}$, κ_B, κ, $Z(\check{U})$, $L(\lambda)^*$, ${}_{\mathcal{B}'}\mathcal{H}_\mathcal{B}$, $\mathcal{C}[P(2\Sigma)]$, Υ, W_Θ, ρ, χ, $g_{2\lambda}$, $\varphi_{2\lambda}$, $\check{U}^B$.

Defined in Section 2:

ω_i	fundamental weight corresponding to α_i
ω_i'	fundamental restricted weight corresponding to $\tilde{\alpha}_i$
W	Weyl group of the root system of $\mathfrak{g}$
t	The rank of Σ

Defined in Section 3:

$w.q^{(\rho,\mu)}\tau(\mu)$	$q^{(\rho,w\mu)}\tau(w\mu)$
$\mathcal{P}$	the Harish-Chandra map defined using (3.3)
$\mathfrak{a}^*$	$\{\tilde{\alpha} \mid \alpha \in \mathfrak{h}^*\}$
$\check{\mathcal{A}}$	$\{\tau(\mu) \mid \mu \in P(\Sigma)\}$
$w \circ q^{(\tilde{\rho},\mu)}\tau(\mu)$	$q^{(\tilde{\rho},w\mu)}\tau(w\mu)$
W'	Weyl group associated to root system of π_Θ
$\check{T}_\Theta$	$\{\tau((\mu+\Theta(\mu))/2) \mid \mu \in P(\pi)\}$
$\tilde{\mathcal{P}}$	projection of $\check{U}^0$ onto $\mathcal{C}[\check{\mathcal{A}}]$ using (3.5)
$\mathcal{A}$	$\{\tau(2\mu) \mid \mu \in P(\Sigma)\}$

Defined in Section 4:

$\mathcal{M}^+$	$\mathcal{M} \cap U^+$
$\mathcal{M}^-$	$\mathcal{M} \cap G^-$
N^+	subalgebra of U^+ generated by $(\mathrm{ad}\ \mathcal{M}^+)\mathcal{C}[x_i \mid \alpha_i \notin \pi_\Theta\}$
N^-	subalgebra of G^- generated by $(\mathrm{ad}\ \mathcal{M}^-)\mathcal{C}[y_i t_i \mid \alpha_i \notin \pi_\Theta]$
$S_{\beta,r}$	the sum of weight spaces $S_{\beta'}$ with $\tilde{\beta}' = \tilde{\beta}$
$T(\gamma)$	the set of $\tau(\eta)$ in T satisfying (4.1)
T'	group generated by t_i for $\alpha_i \in \pi^*$
$\mathcal{P}_B$	see Definition 4.3
G^+	algebra generated by $x_i t_i^{-1}$ for $1 \leq i \leq n$
U^-	algebra generated by y_i for $1 \leq i \leq n$

Defined in Section 5:

$\mathcal{C}[Q(\Sigma)]\mathcal{A}$	ring generated by $\mathcal{C}[Q(\Sigma)]$ and $\mathcal{A}$ subject to (5.1)
$z^\lambda * \tau(\mu)$	$q^{(\lambda,\mu)}z^\lambda$

$\mathcal{C}(Q(\Sigma))\mathcal{A}$	localization of $\mathcal{C}[Q(\Sigma)]\mathcal{A}$ at $\mathcal{C}[Q(\Sigma)]\setminus\{0\}$
$\mathcal{A}_{\geq}$	$\{\tau(\mu) \mid \mu\in Q^+(\Sigma)\cap P(2\Sigma)\}$.
$U_{\geq}$	$U^+G^-\mathcal{A}_{\geq}$
B'	$\chi(B)$
$\mathcal{X}$	the radial component map (see Theorem 5.4)
$\mathcal{C}((Q(\Sigma)))$	the formal Laurent series ring $\mathcal{C}((z^{-\tilde{\alpha}_i} \mid \alpha_i\in\pi^*))$
$\mathcal{C}[[Q^-(\Sigma)]]$	power series ring in the $z^{-\gamma}$ for $\gamma\in Q^+(\Sigma)$
$\mathcal{C}[[Q^-(\Sigma)]]_+$	augmentation ideal of $\mathcal{C}[[Q^-(\Sigma)]]$
$\mathrm{ht}_r\gamma$	the restricted height of γ (see (5.13))
$\mathrm{top}(X)$	highest degree homogeneous term of X (see (5.18))

Defined in Section 6:

$z_{2\mu}$	unique central element in $\tau(2\mu)+(\mathrm{ad}_r\, U_+)\tau(2\mu)$
$\hat{\tau}(\lambda)$	$\sum_{w\in W}\tau(w\lambda)q^{(\rho,w\lambda)}$
$\hat{m}(2\mu)$	$\sum_{\gamma\in W_\Theta\mu}q^{(\tilde{\rho},2\gamma)}\tau(2\gamma)$
$\check{U}_{\mathbf{C}(q)}$	$\mathbf{C}(q)$ algebra generated by $x_i, y_i, 1\leq i\leq n, \check{T}$
A	$\mathbf{C}[q]_{(q-1)}$
$\hat{U}$	A algebra generated by x_i, y_i, $1\leq i\leq n$, $\frac{(t-1)}{(q-1)}$, $t\in\check{T}$
$\mathfrak{a}$	$\{h-\theta(h)\mid h\in\mathfrak{h}\}$
T_2	$\{\tau(2\mu) \mid \mu\in P(\pi)\}$

Defined in Section 7:

ϕ	Hopf algebra automorphism defined by (7.1)
C_i	B_i-s_i
$\mathcal{L}$	see Definition 7.3 and (7.5)
$\mathcal{A}_{\leq}$	$\{\tau(-2\mu) \mid 2\mu\in Q^+(\Sigma)$ and $\mu\in P(\Sigma)\}$

Defined in Section 8:

$\mathfrak{r},\mathfrak{r}^\theta$	proper maximal symmetric pair in $\mathfrak{g},\mathfrak{g}^\theta$ of type AII
$\mu_1,\dots,\mu_t$	simple positive roots of Σ
$\eta_1,\dots,\eta_t$	restricted fundamental weights
η_0	0
η_{t+1}	0
$m(\lambda)$	$\sum_{\gamma\in W_\Theta\lambda}z^\gamma$
Σ_s	$\{\mu_1,\dots,\mu_t\}$
Σ_s'	a subset of $\{\mu_1,\dots,\mu_t\}$
Σ'	root system generated by Σ_s'
W_Θ'	Weyl group of Σ'
λ'	$\lambda'\in P(\Sigma')$ and $(\lambda-\lambda',Q(\Sigma'))=0$
$\mathcal{N}$	$\sum_{\mu_i\in\Sigma_s\setminus\Sigma_s'}\sum_{\gamma\in Q^+(\Sigma)}\mathbf{N}z^{-\mu_i-\gamma}$
$m'(\lambda)$	$\sum_{\gamma\in W_\Theta'\lambda}z^\gamma$
$\hat{M}_k$	see (8.8) and (8.9)
M_k	the image of $\hat{M}_k$ in $\mathcal{C}[P(\Sigma)]$
$\pi_{\mathfrak{r}}$	the set of simple roots of the root system of $\mathfrak{r}$
$W_{\mathfrak{r}}$	the Weyl group of the root system of $\mathfrak{r}$
z_i	$z_i\in Z(\check{U})$ and $\mathcal{P}(z_i)=\hat{\tau}(2\omega_i)$
Z_k	the image of $\mathcal{P}_B(z_k)$ in $\mathcal{C}[P(\Sigma)]$

Z'_k	see the discussion before Theorem 8.13
π_i	$\{\alpha_j \mid 1 \leq j \leq 2i+1\}$
$\mathfrak{g}_i$	semisimple Lie subalgebra of $\mathfrak{g}$ with simple roots π_i
U_i	subalgebra of U equal to $U_q(\mathfrak{g}_i)$
U_0	$\mathcal{C}$
U_t	U
$\check{U}_i$	simply connected quantized enveloping algebra of $\mathfrak{g}_i$
$\check{U}_0$	$\mathcal{C}$
$\mathcal{L}_i$	Map defined in Definition 7.3 and (7.5) for $U_i, U_i \cap B$
$\mathcal{P}_{B\cap U_i}$	Harish-Chandra map for $U_i, U_i \cap B$
$\hat{m}'(2\lambda)$	$\sum_{\gamma \in W'_\Theta \lambda} q^{(\rho,2\gamma)}\tau(2\gamma)$

Defined in Section 9:

$\mathrm{Supp}(\sum_\gamma a_\gamma \tau(\gamma))$	$\{\tau(\gamma) \mid a_\gamma \neq 0\}$
$c_{2\eta}$	$\mathcal{L}(\tau(2\eta) + b_\eta)$

Bibliography

[A] S. Araki, *On root systems and an infinitesimal classification of irreducible symmetric spaces*, Journal of Mathematics, Osaka City University **13** (1962), no. 1, 1-34.

[B] P. Baumann, *On the center of quantized enveloping algebras*, Journal of Algebra **203** (1998), 244-260.

[CP] V. Chari and A. Pressley, *A Guide to Quantum Groups*, Cambridge University Press, Cambridge, (1995).

[DN] M.S. Dijkhuizen and M. Noumi, *A family of quantum projective spaces and related q-hypergeometric orthogonal polynomials*, Transactions of the American Mathematical Society **350** (1998), no. 8, 3269-3296.

[D] J. Dixmier, *Algèbres Enveloppantes*, Cahiers Scientifiques, XXXVII, Gauthier-Villars, Paris (1974).

[DK] C. DeConcini and V.G. Kac, *Representations of quantum groups at roots of 1*, In: *Operator Algebras, Unitary Representations, Enveloping Algebras, and Invariant Theory*, Progress in Math. **92**, Birkhäuser, Boston (1990), 471-506.

[DS] M.S. Dikhhuizen and J.V. Stokman, *Some limit transitions between BC type orthogonal polynomials interpreted on quantum complex Grassmannians*, Publ. Res. Inst. Math. Sci. **35** (1999), 451-500.

[H] J.E. Humphreys, *Introduction to Lie Algebras and Representation Theory*, Springer-Verlag, New York (1972).

[He] S. Helgason, *Some results on invariant differential operators on symmetric spaces*, American Journal of Mathematics **114**, No. 4, 789-811.

[HC] Harish-Chandra, *Spherical functions on a semisimple Lie group* I, American Journal of Mathematics **80** (1958) 241-310.

[Ja] N. Jacobson, *Basic Algebra* I, W. H. Freeman and Company, San Francisco (1974).

[JL1] A. Joseph and G. Letzter, *Local finiteness of the adjoint action for quantized enveloping algebras*, Journal of Algebra **153** (1992), 289 -318.

[JL2] A. Joseph and G. Letzter,*Separation of variables for quantized enveloping algebras*, American Journal of Mathematics **116** (1994), 127-177.

[Jo] A. Joseph, *Quantum Groups and Their Primitive Ideals*, Springer-Verlag, New York (1995).

[Ke] M.S. Kébé, *$\mathcal{O}$-algèbres quantiques*, C. R. Académie des Sciences, Série I, Mathématique **322** (1996), no. 1, 1-4.

[K] A.A. Kirillov, Jr.,*Lectures on affine Hecke algebras and Macdonald's conjectures*, Bulletin of the American Mathematical Society **34** (1997), No. 3, 251-292.

[KS] E. Koelink, J. Stokman, *Fourier transforms on the quantum SU(1,1) group. With an appendix by Mizan Rahman* Publications of the Kyoto University Research Institute for Mathematical Science **37** (2001), no. 4, 621-715.

[Kn] A. W. Knapp, *Lie Groups Beyond an Introduction*, Progress in Math. **140**, Birkhäuser, Boston (1996).

[Le] J. Lepowsky, *On the Harish-Chandra Homomorphism*, Transactions of the American Mathematical Society **208** (1975) 193-218.

[L1] G. Letzter, *Symmetric pairs for quantized enveloping algebras*, Journal of Algebra **220** (1999), no. 2, 729-767.

[L2] G. Letzter, *Harish-Chandra modules for quantum symmetric pairs*, Representation Theory, An Electronic Journal of the American Mathematical Society **4** (1999) 64-96.

[L3] G. Letzter, *Coideal subalgebras and quantum symmetric pairs*, In: *New Directions in Hopf Algebras, Mathematical Sciences Research Institute Publications* **43**, Cambridge University Press (2002), 117-166.

[L4] G. Letzter, *Quantum symmetric pairs and their zonal spherical functions*, Transformation Groups **8** (2003), no. 3, 261-292.

[L5] G. Letzter, *Quantum zonal spherical functions and Macdonald polynomials*, Advances in Mathematics **189** (2004), no. 1, 88 - 147.

[M] I.G. Macdonald, *Orthogonal polynomials associated with root systems*, Séminaire Lotharingien de Combinatoire **45** (2000/01) 40 pp.

[N] M. Noumi,*Macdonald's symmetric polynomials as zonal spherical functions on some quantum homogeneous spaces*, Advances in Mathematics **123** (1996), no. 1, 16-77.

[NDS] M. Noumi, M.S. Dijkhuizen, and T. Sugitani, *Multivariable Askey-Wilson polynomials and quantum complex Grassmannians*, Fields Institute Communications **14** (1997), 167-177.

[NS] M. Noumi and T. Sugitani, *Quantum symmetric spaces and related q-orthogonal polynomials*, In: *Group Theoretical Methods in Physics (ICGTMP)* (Toyonaka, Japan, 1994), World Science Publishing, River Edge, New Jersey (1995), 28-40.

[R] M. Rosso, Groupes Quantiques, *Représentations linéaires et applications, Thèse*, Paris 7 (1990).

[S] T. Sugitani, *Zonal spherical functions on quantum Grassmann manifolds*, The University of Tokyo Journal of Mathematical Sciences **6** (1999), no. 2, 335-369.

Editorial Information

To be published in the *Memoirs*, a paper must be correct, new, nontrivial, and significant. Further, it must be well written and of interest to a substantial number of mathematicians. Piecemeal results, such as an inconclusive step toward an unproved major theorem or a minor variation on a known result, are in general not acceptable for publication.

Papers appearing in *Memoirs* are generally at least 80 and not more than 200 published pages in length. Papers less than 80 or more than 200 published pages require the approval of the Managing Editor of the Transactions/Memoirs Editorial Board.

As of January 31, 2008, the backlog for this journal was approximately 17 volumes. This estimate is the result of dividing the number of manuscripts for this journal in the Providence office that have not yet gone to the printer on the above date by the average number of monographs per volume over the previous twelve months, reduced by the number of volumes published in four months (the time necessary for preparing a volume for the printer). (There are 6 volumes per year, each usually containing at least 4 numbers.)

A Consent to Publish and Copyright Agreement is required before a paper will be published in the *Memoirs*. After a paper is accepted for publication, the Providence office will send a Consent to Publish and Copyright Agreement to all authors of the paper. By submitting a paper to the *Memoirs*, authors certify that the results have not been submitted to nor are they under consideration for publication by another journal, conference proceedings, or similar publication.

Information for Authors

Memoirs are printed from camera copy fully prepared by the author. This means that the finished book will look exactly like the copy submitted.

Initial submission. The AMS uses Centralized Manuscript Processing for initial submissions. Authors should submit a PDF file using the Initial Manuscript Submission form found at `www.ams.org/cgi-bin/peertrack/submission.pl`, or send one copy of the manuscript to the following address: Centralized Manuscript Processing, MEMOIRS OF THE AMS, 201 Charles Street, Providence, RI 02904-2294 USA. If a paper copy is being forwarded to the AMS, indicate that it is for it Memoirs and include the name of the corresponding author, contact information such as email address or mailing address, and the name of an appropriate Editor to review the paper (see the list of Editors below).

The paper must contain a *descriptive title* and an *abstract* that summarizes the article in language suitable for workers in the general field (algebra, analysis, etc.). The *descriptive title* should be short, but informative; useless or vague phrases such as "some remarks about" or "concerning" should be avoided. The *abstract* should be at least one complete sentence, and at most 300 words. Included with the footnotes to the paper should be the 2000 *Mathematics Subject Classification* representing the primary and secondary subjects of the article. The classifications are accessible from `www.ams.org/msc/`. The list of classifications is also available in print starting with the 1999 annual index of *Mathematical Reviews*. The Mathematics Subject Classification footnote may be followed by a list of *key words and phrases* describing the subject matter of the article and taken from it. Journal abbreviations used in bibliographies are listed in the latest *Mathematical Reviews* annual index. The series abbreviations are also accessible from `www.ams.org/publications/`. To help in preparing and verifying references, the AMS offers MR Lookup, a Reference Tool for Linking, at `www.ams.org/mrlookup/`.

Electronically prepared manuscripts. The AMS encourages electronically prepared manuscripts, with a strong preference for $\mathcal{A}\mathcal{M}\mathcal{S}$-LaTeX. To this end, the Society has prepared $\mathcal{A}\mathcal{M}\mathcal{S}$-LaTeX author packages for each AMS publication. Author packages include instructions for preparing electronic manuscripts, samples, and a style file that generates the particular design specifications of that publication series. Though $\mathcal{A}_{\mathcal{M}}\mathcal{S}$-LaTeX is the highly preferred format of TeX, author packages are also available in $\mathcal{A}_{\mathcal{M}}\mathcal{S}$-TeX.

Authors may retrieve an author package from the AMS website starting from `www.ams.org/tex/` or via FTP to `ftp.ams.org` (login as `anonymous`, enter username as password, and type `cd pub/author-info`). The *AMS Author Handbook* and the *Instruction Manual* are available in PDF format following the author packages link from `www.ams.org/tex/`. The author package can also be obtained free of charge by sending email to `tech-support@ams.org` (Internet) or from the Publication Division, American Mathematical Society, 201 Charles St., Providence, RI 02904-2294, USA. When requesting an author package, please specify $\mathcal{A}_{\mathcal{M}}\mathcal{S}$-LaTeX or $\mathcal{A}_{\mathcal{M}}\mathcal{S}$-TeX and the publication in which your paper will appear. Please be sure to include your complete mailing address.

After acceptance. The final version of the electronic file should be sent to the Providence office (this includes any TeX source file, any graphics files, and the DVI or PostScript file) immediately after the paper has been accepted for publication.

Before sending the source file, be sure you have proofread your paper carefully. The files you send must be the EXACT files used to generate the proof copy that was accepted for publication. For all publications, authors are required to send a printed copy of their paper, which exactly matches the copy approved for publication, along with any graphics that will appear in the paper.

Accepted electronically prepared files can be submitted via the web at `www.ams.org/submit-book-journal/`, sent via FTP, or sent on CD-Rom or diskette to the Electronic Prepress Department, American Mathematical Society, 201 Charles Street, Providence, RI 02904-2294 USA. TeX source files, DVI files, and PostScript files can be transferred over the Internet by FTP to the Internet node `ftp.ams.org` (130.44.1.100). When sending a manuscript electronically via CD-Rom or diskette, please be sure to include a message identifying the paper as a Memoir.

Electronically prepared manuscripts can also be sent via email to `pub-submit@ams.org` (Internet). In order to send files via email, they must be encoded properly. (DVI files are binary and PostScript files tend to be very large.)

Electronic graphics. Comprehensive instructions on preparing graphics are available at `www.ams.org/jourhtml/`. A few of the major requirements are given here.

Submit files for graphics as EPS (Encapsulated PostScript) files. This includes graphics originated via a graphics application as well as scanned photographs or other computer-generated images. If this is not possible, TIFF files are acceptable as long as they can be opened in Adobe Photoshop or Illustrator. No matter what method was used to produce the graphic, it is necessary to provide a paper copy to the AMS.

Authors using graphics packages for the creation of electronic art should also avoid the use of any lines thinner than 0.5 points in width. Many graphics packages allow the user to specify a "hairline" for a very thin line. Hairlines often look acceptable when proofed on a typical laser printer. However, when produced on a high-resolution laser imagesetter, hairlines become nearly invisible and will be lost entirely in the final printing process.

Screens should be set to values between 15% and 85%. Screens which fall outside of this range are too light or too dark to print correctly. Variations of screens within a graphic should be no less than 10%.

Inquiries. Any inquiries concerning a paper that has been accepted for publication should be sent to `memo-query@ams.org` or directly to the Electronic Prepress Department, American Mathematical Society, 201 Charles St., Providence, RI 02904-2294 USA.

Editors

This journal is designed particularly for long research papers, normally at least 80 pages in length, and groups of cognate papers in pure and applied mathematics. Papers intended for publication in the *Memoirs* should be addressed to one of the following editors. The AMS uses Centralized Manuscript Processing for initial submissions to AMS journals. Authors should follow instructions listed on the Initial Submission page found at www.ams.org/memo/memosubmit.html.

Algebra to ALEXANDER KLESHCHEV, Department of Mathematics, University of Oregon, Eugene, OR 97403-1222; email: ams@noether.uoregon.edu

Algebraic geometry and its application to MINA TEICHER, Emmy Noether Research Institute for Mathematics, Bar-Ilan University, Ramat-Gan 52900, Israel; email: teicher@macs.biu.ac.il

Algebraic geometry to DAN ABRAMOVICH, Department of Mathematics, Brown University, Box 1917, Providence, RI 02912; email: amsedit@math.brown.edu

Algebraic number theory to V. KUMAR MURTY, Department of Mathematics, University of Toronto, 100 St. George Street, Toronto, ON M5S 1A1, Canada; email: murty@math.toronto.edu

Algebraic topology to ALEJANDRO ADEM, Department of Mathematics, University of British Columbia, Room 121, 1984 Mathematics Road, Vancouver, British Columbia, Canada V6T 1Z2; email: adem@math.ubc.ca

Combinatorics to JOHN R. STEMBRIDGE, Department of Mathematics, University of Michigan, Ann Arbor, Michigan 48109-1109; email: FRS@umich.edu

Complex analysis and harmonic analysis to ALEXANDER NAGEL, Department of Mathematics, University of Wisconsin, 480 Lincoln Drive, Madison, WI 53706-1313; email: nagel@math.wisc.edu

Differential geometry and global analysis to LISA C. JEFFREY, Department of Mathematics, University of Toronto, 100 St. George St., Toronto, ON Canada M5S 3G3; email: jeffrey@math.toronto.edu

Functional analysis and operator algebras to DIMITRI SHLYAKHTENKO, Department of Mathematics, University of California, Los Angeles, CA 90095; email: shlyakht@math.ucla.edu

Geometric analysis to WILLIAM P. MINICOZZI II, Department of Mathematics, Johns Hopkins University, 3400 N. Charles St., Baltimore, MD 21218; email: trans@math.jhu.edu

Geometric analysis to MARK FEIGHN, Math Department, Rutgers University, Newark, NJ 07102; email: feighn@andromeda.rutgers.edu

Harmonic analysis, representation theory, and Lie theory to ROBERT J. STANTON, Department of Mathematics, The Ohio State University, 231 West 18th Avenue, Columbus, OH 43210-1174; email: stanton@math.ohio-state.edu

Logic to STEFFEN LEMPP, Department of Mathematics, University of Wisconsin, 480 Lincoln Drive, Madison, Wisconsin 53706-1388; email: lempp@math.wisc.edu

Number theory to JONATHAN ROGAWSKI, Department of Mathematics, University of California, Los Angeles, CA 90095; email: jonr@math.ucla.edu

Partial differential equations to GUSTAVO PONCE, Department of Mathematics, South Hall, Room 6607, University of California, Santa Barbara, CA 93106; email: ponce@math.ucsb.edu

Partial differential equations and dynamical systems to PETER POLACIK, School of Mathematics, University of Minnesota, Minneapolis, MN 55455; email: polacik@math.umn.edu

Probability and statistics to RICHARD BASS, Department of Mathematics, University of Connecticut, Storrs, CT 06269-3009; email: bass@math.uconn.edu

Real analysis and partial differential equations to DANIEL TATARU, Department of Mathematics, University of California, Berkeley, Berkeley, CA 94720; email: tataru@math.berkeley.edu

All other communications to the editors should be addressed to the Managing Editor, ROBERT GURALNICK, Department of Mathematics, University of Southern California, Los Angeles, CA 90089-1113; email: guralnic@math.usc.edu.

Titles in This Series

905 **Dominic Verity,** Complicial sets characterising the simplicial nerves of strict ω-categories, 2008
904 **William M. Goldman and Eugene Z. Xia,** Rank one Higgs bundles and representations of fundamental groups of Riemann surfaces, 2008
903 **Gail Letzter,** Invariant differential operators for quantum symmetric spaces, 2008
902 **Bertrand Toën and Gabriele Vezzosi,** Homotopical algebraic geometry II: Geometric stacks and applications, 2008
901 **Ron Donagi and Tony Pantev (with an appendix by Dmitry Arinkin),** Torus fibrations, gerbes, and duality, 2008
900 **Wolfgang Bertram,** Differential geometry, Lie groups and symmetric spaces over general base fields and rings, 2008
899 **Piotr Hajłasz, Tadeusz Iwaniec, Jan Malý, and Jani Onninen,** Weakly differentiable mappings between manifolds, 2008
898 **John Rognes,** Galois extensions of structured ring spectra/Stably dualizable groups, 2008
897 **Michael I. Ganzburg,** Limit theorems of polynomial approximation with exponential weights, 2008
896 **Michael Kapovich, Bernhard Leeb, and John J. Millson,** The generalized triangle inequalities in symmetric spaces and buildings with applications to algebra, 2008
895 **Steffen Roch,** Finite sections of band-dominated operators, 2008
894 **Martin Dindoš,** Hardy spaces and potential theory on C^1 domains in Riemannian manifolds, 2008
893 **Tadeusz Iwaniec and Gaven Martin,** The Beltrami Equation, 2008
892 **Jim Agler, John Harland, and Benjamin J. Raphael,** Classical function theory, operator dilation theory, and machine computation on multiply-connected domains, 2008
891 **John H. Hubbard and Peter Papadopol,** Newton's method applied to two quadratic equations in $\mathbb{C}^2$ viewed as a global dynamical system, 2008
890 **Steven Dale Cutkosky,** Toroidalization of dominant morphisms of 3-folds, 2007
889 **Michael Sever,** Distribution solutions of nonlinear systems of conservation laws, 2007
888 **Roger Chalkley,** Basic global relative invariants for nonlinear differential equations, 2007
887 **Charlotte Wahl,** Noncommutative Maslov index and eta-forms, 2007
886 **Robert M. Guralnick and John Shareshian,** Symmetric and alternating groups as monodromy groups of Riemann surfaces I: Generic covers and covers with many branch points, 2007
885 **Jae Choon Cha,** The structure of the rational concordance group of knots, 2007
884 **Dan Haran, Moshe Jarden, and Florian Pop,** Projective group structures as absolute Galois structures with block approximation, 2007
883 **Apostolos Beligiannis and Idun Reiten,** Homological and homotopical aspects of torsion theories, 2007
882 **Lars Inge Hedberg and Yuri Netrusov,** An axiomatic approach to function spaces, spec tral synthesis and Luzin approximation, 2007
881 **Tao Mei,** Operator valued Hardy spaces, 2007
880 **Bruce C. Berndt, Geumlan Choi, Youn-Seo Choi, Heekyoung Hahn, Boon Pin Yeap, Ae Ja Yee, Hamza Yesilyurt, and Jinhee Yi,** Ramanujan's forty identities for Rogers-Ramanujan functions, 2007
879 **O. García-Prada, P. B. Gothen, and V. Muñoz,** Betti numbers of the moduli space of rank 3 parabolic Higgs bundles, 2007
878 **Alessandra Celletti and Luigi Chierchia,** KAM stability and celestial mechanics, 2007
877 **María J. Carro, José A. Raposo, and Javier Soria,** Recent developments in the theory of Lorentz spaces and weighted inequalities, 2007

876 **Gabriel Debs and Jean Saint Raymond,** Borel liftings of Borel sets: Some decidable and undecidable statements, 2007

875 **C. Krattenthaler and T. Rivoal,** Hypergéométrie et fonction zêta de Riemann, 2007

874 **Sonia Natale,** Semisolvability of semisimple Hopf algebras of low dimension, 2007

873 **A. J. Duncan,** Exponential genus problems in one-relator products of groups, 2007

872 **Anthony V. Geramita, Tadahito Harima, Juan C. Migliore, and Yong Su Shin,** The Hilbert function of a level algebra, 2007

871 **Pascal Auscher,** On necessary and sufficient conditions for L^p-estimates of Riesz transforms associated to elliptic operators on $\mathbb{R}^n$ and related estimates, 2007

870 **Takuro Mochizuki,** Asymptotic behaviour of tame harmonic bundles and an application to pure twistor D-modules, Part 2, 2007

869 **Takuro Mochizuki,** Asymptotic behaviour of tame harmonic bundles and an application to pure twistor D-modules, Part 1, 2007

868 **Gelu Popescu,** Entropy and multivariable interpolation, 2006

867 **Vilmos Totik,** Metric properties of harmonic measures, 2006

866 **William Craig,** Semigroups underlying first-order logic, 2006

865 **Nathanial P. Brown,** Invariant means and finite representation theory of $C*$-algebras, 2006

864 **John M. Lee,** Fredholm operators and Einstein metrics on conformally compact manifolds, 2006

863 **M. Lübke and A. Teleman,** The Universal Kobayashi-Hitchin correspondence on Hermitian manifolds, 2006

862 **Alberto Canonaco,** The Beilinson complex and canonical rings of irregular surfaces, 2006

861 **Leon A. Takhtajan and Lee-Peng Teo,** Weil-Petersson metric on the universal Teichmüller space, 2006

860 **Thomas M. Fiore,** Pseudo limits, biadjoints and pseudo algebras: Categorical foundations of conformal field theory, 2006

859 **N. Arcozzi, R. Rochberg, and E. Sawyer,** Carleson measures and interpolating sequences for Besov spaces on complex balls, 2006

858 **Enrico Valdinoci, Berardino Sciunzi, and Vasile Ovidiu Savin,** Flat level set regularity of p-Laplace phase transitions, 2006

857 **Donatella Danielli, Nocola Garofalo, and Duy-Minh Nhieu,** Non-doubling Ahlfors measures, perimeter measures, and the characterization of the trace spaces of Sobolev functions in Carnot-Carathéodory spaces, 2006

856 **Vladimir Bolotnikov and Harry Dym,** On boundary interpolation for matrix valued Schur functions, 2006

855 **Yevgenia Kashina, Yorck Sommerhäuser, and Yongchang Zhu,** On higher Frobenius-Schur indicators, 2006

854 **Noam Greenberg,** The role of true finiteness in the admissible recursively enumerable degrees, 2006

853 **Joachim Krieger,** Stability of spherically symmetric wave maps, 2006

852 **Viorel Barbu, Irena Lasiecka, and Roberto Triggiani,** Tangential boundary stabilization of Navier-Stokes equations, 2006

851 **Jie Wu,** On maps from loop suspensions to loop spaces and the shuffle relations on the Cohen groups, 2006

For a complete list of titles in this series, visit the
AMS Bookstore at **www.ams.org/bookstore/**.